Lecture Notes in Comput 1

Commenced Publication in 1973
Founding and Former Series Editors:
Gerhard Goos, Juris Hartmanis, and Jan va.

Thomas Magedanz   Edmundo R.M. Madeira
Petre Dini (Eds.)

Operations
and Management
in IP-Based Networks

5th IEEE International Workshop on
IP Operations and Management, IPOM 2005
Barcelona, Spain, October 26-28, 2005
Proceedings

 Springer

Volume Editors

Thomas Magedanz
Technical University of Berlin/Fraunhofer Institute FOKUS
Kaiserin-Augusta-Allee 31, 10589 Berlin, Germany
E-mail: magedanz@fokus.fraunhofer.de

Edmundo R.M. Madeira
UNICAMP – University of Campinas, IC - Institute of Computing
13083-970 Campinas, SP, Brazil
E-mail: edmundo@ic.unicamp.br

Petre Dini
Cisco Systems, Inc.
10 West Tasman Drive, Bldg O/2, San Jose, CA 95134, USA
E-mail: pdini@cisco.com

Library of Congress Control Number: 2005933267

CR Subject Classification (1998): C.2, D.4.4, D.2, H.3.5, H.4, K.6.4

ISSN   0302-9743
ISBN-10   3-540-29356-6 Springer Berlin Heidelberg New York
ISBN-13   978-3-540-29356-9 Springer Berlin Heidelberg New York

Springer is a part of Springer Science+Business Media

springeronline.com

© Springer-Verlag Berlin Heidelberg 2005
Printed in Germany

Typesetting: Camera-ready by author, data conversion by Olgun Computergrafik
Printed on acid-free paper   SPIN: 11567486   06/3142   5 4 3 2 1 0

Preface

The beginning of the 21st century is witnessing a drive to the convergence of fixed and mobile telecommunication networks and the increasing adoption of IP technologies for implementing seamless multimedia applications in next-generation networks. The IEEE International Workshop Series on IP Operations & Management (IPOM) is documenting this evolution by providing snapshots of the state of the art in the field of operations and management in IP-based networks.

The 5th IEEE International Workshop on IP Operations & Management (IPOM 2005), devoted to the "O&M Challenges in Next Generation Services and Networks", was held in Barcelona, Spain, October 26–28, 2005. Here IPOM was one of the five collocated events under the banner "First International Week on Management on Networks and Services (www.manweek2005.org)", together with the 16th IFIP/IEEE International Workshop on Distributed Systems: Operations and Management (DSOM 2005), the 8th International Conference on Management of Multimedia Networks and Services (MMNS 2005), the 2005 Symposium on Self-stabilizing Systems (SSS 2005) and the 1st IEEE/IFIP International Workshop on Autonomic Grid Networking and Management (AGNM 2005).

This book contains the official proceedings of IPOM 2005. It features 21 high-quality papers grouped into seven technical sessions looking at O&M for VoIP, IMS and managed IP services, management of open interfaces, QoS and pricing in NGNs, autonomic communications, policy-based management, routing and topologies, routing and tools, as well as experiences from testbeds and trials. Additional papers presented in two short sessions are published separately.

We would like to thank the authors for all their efforts, as well as the members of the Technical Program Committee, and the reviewers. Without their support the high-quality program of this event would not have been possible. We are also indebted to many individuals and organizations that made the conference possible (IEEE, IARIA, Fraunhofer Institute FOKUS, JEMS's drivers, and the Universitat Politècnica de Catalunya).

October 2005

Thomas Magedanz
Edmundo R.M. Madeira
Petre Dini

Organization

General Chair

Petre Dini, Cisco Systems, USA

Program Co-chairs

Thomas Magedanz, TU Berlin / Fraunhofer FOKUS, Germany
Edmundo Madeira, Unicamp, Brazil

Technical Program Committee

Alexander Clemm, Cisco Systems, USA
Andrzej Jajszczyk, AGH University of Science and Technology, Poland
Antonio Pescapé, University of Napoli "Federico II", Italy
Carlos Becker Westphall, UFSC, Brazil
Deep Medhi, University of Missouri-Kansas City, USA
G.-S. Kuo, NCCU, Republic of China
Gerard Parr, University of Ulster, UK
Iakovos Venieris, NTUA, Greece
Iraj Saniee, Bell Labs, USA
Joerg Heuer, T-Labs, Germany
Johan Zuidweg, Telefonica, Spain
John-Luc Bakker, Telcordia, USA
Karthikeyan Ramasamy, Juniper, USA
Manu Malek, Stevens Institute of Technology, USA
Marcus Brunner, NEC Europe, Germany
Mario Baldi, Politecnico di Torino, Italy
Martin Stiemerling, NEC, Germany
Masum Hasan, Cisco Systems, USA
Meng Luoming, BUPT, P.R. China
Michal Pioro, Warsaw University of Technology, Poland
Nail Akar, Bilkent University, Turkey
Peter Domschitz, Alcatel, Germany
Richard Schaedler, Tekelec, USA
Roberto Minerva, Telecom Italy Labs, Italy
Sascha Karlich, Siemens, Austria
Stamatios Kartalopoulos, University of Oklahoma, USA
Tassos Gavras, EURESCOM, Germany
Tom Chen, SMU, USA
William Donnelly, Waterford Institute of Technology, Ireland
Wolfgang Kellerer, DoCoMo Eurolabs, Germany
Wulf Bauerfeld, T-Systems, Germany

Panel Chair

Stephan Steglich, FOKUS, Germany

Steering Committee

Tom Chen, Southern Methodist University, USA
Petre Dini, Cisco Systems, USA
Andrzej Jajszczyk, AGH University of Science and Technology, Poland
G.-S. Kuo, NCCU, Republic of China
Deep Medhi, University of Missouri-Kansas City, USA
Curtis Siller, IEEE ComSoc, USA

Table of Contents

Routing

Routing and Tools

Experiences from Testbeds and Trials

Emergency Telecommunication Support
for IP Telephony

Francesco Moggia, Mudumbai Ranganathan, Eunsook Kim, and Doug Montgomery

Advanced Networking Technologies Division,
National Institute of Standards and Technology,
Gaithersburg, MD 20899, USA
{fmoggia,mranga,eunah,dougm}@nist.gov
http://w3.antd.nist.gov

Abstract. As universal high speed internet access becomes a reality, phone calls are increasingly being made over the Internet rather than the conventional PSTN. The danger to this trend is the un-availability of priority mechanisms for communication between emergency response personnel during times of disaster. We define a proposed architecture to enable ETS support for SIP-based VOIP systems.

1 Introduction

The abundance of available bandwidth has spurred the growth of companies that provide telephony services over the public Internet at minimal cost. This has prompted many to abandon the Public Switched Telephone Network (PSTN) altogether and rely solely upon IP Telephony. IP Telephony offers some advantages for disaster response. For example, one may directly contact the Public Safety Access Point (PSAP) of a region by specifying the region in a URL or as URL parameters. The personal mobility offered by such technologies permits people to contact each other regardless of physical location. Integration with web technologies and enhanced methods of communication also permit richer interactions between emergency response personnel, resulting in an improved ability to react to the emergency scenario at hand. However there are some negative aspects to IP Telephony under such circumstances. Recent experience with disaster scenarios indicate that the public telephone network will become so overloaded that effective communication is no longer possible. Similarly, for IP telephony, without dedicated capacity, it is expected that an overload of the signaling servers and media gateways will occur.

The PSTN is regulated by Government and providers are required to allocate dedicated resources for emergency calling – allowing authorized personnel to complete even when the normal (unreserved) network is fully saturated. However, the same cannot be guaranteed for telephony over the Internet. Service providers are generally reluctant to accept any form of government regulation. How do we provide authorized personnel with priority calling under these circumstances? In this paper we outline a solution to this problem by proposing a coupling of signaling and Quality of Service (QOS) mechanisms for IP telephony.

There are two essential parts to IP Telephony – signaling or call setup and media (RTP). To devise an effective solution for emergency telecommunications both signaling and media issues should be addressed. During disasters, signaling services

T. Magedanz, E.R.M. Madeira, and P. Dini (Eds.): IPOM 2005, LNCS 3751, pp. 1–8, 2005.

may become saturated and media gateways may suffer overloads. Emergency calls may need to traverse and Broadband access networks and sections of the Internet located in regions of high congestion to reach other emergency workers. In this paper we make the assumption that QOS mechanism such as DIFFSERV is deployed in the network and propose architecture based on this assumption.

The rest of this paper is organized as follows. In section 2 we briefly outline the Emergency scenario to set the stage for the issues involved in contrast with how these issues are addressed in the PSTN. In section 3, we discuss these issues in greater detail. In section 4 we outline the architectural assumptions of our proposed solution. In section 5 and 6 we detail or proposed solution.

2   The Emergency Scenario

The requirements for IP Emergency Telecommunication systems are outlined in RFC3690 When a public disaster occurs, emergency workers need to communicate with other emergency workers outside the disaster zone for effective coordination. Under typical circumstances they will be using the same telephony signaling servers and media gateways that are being used by regular phone users, thus resulting in an overload of these.

The PSTN has government mandated support for emergency calling. This service is known as GETS (Government Emergency Telecommunications Support). A GETS user is issued a special card that can be used from any public telephone. This gives the user a higher probability of call completion by dedicating resources to the call. Thus even when regular users are getting busy signals or "all circuits are busy" messages, the GETS user gets through. GETS support on PSTN relies on (1) the availability of alternate paths (2) preferential treatment at Signaling Control Points (SCPs) – a GETS call setup message is less likely to be dropped in an SS7 network. Similar to GETS, the Wireless Priority Service (WPS) provides resource prioritization for emergency call setup for wireless technologies. There is no equivalent emergency call setup mechanism over the IP network today.

3   Issues in Supporting ETS in IP Telephony

Since the PSTN is a managed network end-to-end measures that give priority to GETS calls are possible. Unlike the PSTN, the IP network uses dynamic routing of packets and the packet may traverse multiple IP service provider networks. A call may not be IP End to End; a call may traverse multiple segments – some of which are circuit switched and others of which are packet networks. Each such switching point will have a gateway that supports a fixed number of trunk lines. Some sections of the network may thus be supported by GETS/WPS and others may not. Such a hybrid scenario is in Figure 1.

During call setup, when the call transitions from IP to PSTN, the signaling server communicates with a server to potentially reserve a trunk line for the call. The signaling and media for an end-to-end call may in fact traverse several such gateways. When a given call transition between the PSTN and Internet or cellular network, there is a high probability of trunk blocking caused by the limited number of trunk lines supported by the gateway. It may thus become necessary to pre-empt trunk lines for use by emergency workers or reserve trunk lines for emergency use. The challenges at hand are (1) Authentication and authorization for the use of ETS over IP networks. (2) Agreed upon service policies with service providers (3) Identification of Emergency connection requests (4) Priority treatment call setup signaling and media (5) Reservation of resources for emergency calls (6) Priority reservation of trunks and bandwidth resources under emergency conditions.

In this paper, we restrict our attention to an end-to-end IP based solution and consider the support requirements of the access point to the IP network.

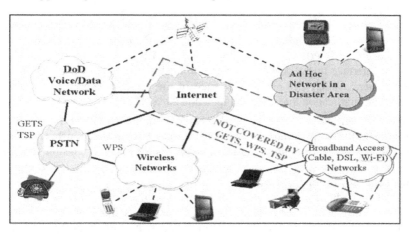

Fig. 1. Emergency Preparedness – GETS, WPS and TSP. An emergency call may have to traverse different networks operated by different providers

4   ETS Support of IP Telephony – Architectural Assumptions

For the purposes of this paper, we assume that SIP is the signaling protocol for call setup. There are two sub-problems that need to be addressed – (1) signaling prioritization for high priority calls and (2) preferential handling of media packets for emergency workers to communicate in a congested network. The main issues in achieving these objectives are (1) mechanisms resource prioritization at the service platform for preferential treatment of the call and (2) guarding against masquerading by regular users to get better quality of service. The main focus of this paper is to define an architecture to achieve these objectives.

5   Supporting Resource Prioritization in a Service Platform

Services are fragments of code that intervene during the signaling for call setup. The Service Container or Service platform is a managed, middleware environment where these fragments of code are installed and execute in a contained environment. The Service Plane is where the network intelligence resides. Standards are just being defined for service architectures at present. The JAIN Service Logic Execution Environment (JSLEE) [3] is an emerging JAVA standard that is designed explicitly for high throughput, low latency asynchronous applications such as IP Telephony Services.

5.1 Architectural Overview of the Service Platform

To experiment with ideas on how to support ETS in managed networks, we picked JSLEE as a concrete standard because it offers a complete set of features and facilities. We now give a brief description of JSLEE in this section and outline the enhancements to support Signaling Prioritization.

The JSLEE application server is composed of three main parts:

- The Execution Logic: this is the core of the SLEE that runs the deployed application
- The Resource Adaptor: Resources are the way in which a SLEE application can interact with the outside world (Protocol Stacks, Databases, Network Devices)
- Management Interface and Facilities: The JSLEE standard provides a Management Interface that allows the system administrator to deploy a service inside the container, manage the operational state of the service, receive alarm and trace from the application that are running inside the SLEE.

A JSLEE Service is composed of elementary components called Service Building Blocks (SBB) that are composed together in a parent-child relationship with execution priorities that defines their execution order. The SLEE is responsible for instantiating services and instances of the components of a service and routing events to the components at run time. JSLEE follows a Publish Subscribe model i.e. applications are bundled with deployment descriptors that specify the conditions under which component instances receive events and the conditions under which Services are instantiated. A SBB is essentially an event handler. It is executed by the container when an event that it is interested in is received. A Service is a composition of these event handlers. The SLEE manages the instantiation and destruction of the SBBs in addition to routing events to it. A Service developer defines the conditions under which the Service is instantiated and installs a template of the service using a deployment descriptor. A Service instance can be seen as a tree composed of SBB instances. Event routing and priority is controlled by the parent child relationship defined in the Service deployment descriptor, which determines the delivery order of the events to the SBBs that are part of the same service.

JSLEE defines an abstraction for an event bus. Events are logically grouped and fired in FIFO order on an event bus or Activity by a *Resource Adaptor* (protocol stack). This is mapped by the SLEE to an internal data structure known as an *ActivityContext*. The ActivityContext has a corresponding application-visible data structure called an *ActivityContextInterface*. Protocol stacks (such as a SIP protocol stack) are pluggable *Resource Adaptors*. The primary function of such a Resource Adaptor is to identify streams of events and place the event on the SLEE abstraction for the related stream of events (i.e. the ActivityContext). The SLEE's primary job at runtime is to identify services that may need to be instantiated on receipt of these events and route events to the SBBs of such services (which may already be instantiated previously).

As a concrete example, the Java Call Control (JCC) Resource Adaptor may regard a phone call as an Activity with an accompanying ActivityContext and identify the incoming signaling messages related to the call as belonging to that Activity (in the case of SIP, it would use the SIP Call ID to do so). In order to receive events related to an Activity an SBB should be attached to the corresponding ActivityContext. This attachment is done by the SLEE runtime environment the SLEE Event Router.

5.2   Supporting Resource Prioritization in JAIN-SLEE

The Signaling server can become the choke point during times of high utilization. During busy periods, it can be expected to handle thousands of call attempts per second. Hence it is important to have mechanisms to prioritize CPU use and have mechanisms by means of which high priority signaling events get preferential treatment during times of emergency.

We propose mechanisms to deal with this. These mechanisms are geared towards increasing the platform resources available to handle calls from emergency response personnel and rely on the decoupled publish-subscribe model and the component model of the SLEE.

When a Service is specified, it is defined by the user as a set of Abstract Classes which comprise its SBBs and a deployment descriptor that ties these together. When the SLEE installs the Service are converted by the SLEE Deployable Unit loader to concrete Java classes and each SBB is assigned its own Java class-loader, thus providing environment isolation. This allows us the opportunity to insert monitoring code fragments using bytecode rewriting techniques to ensure that an SBB cannot exceed allocated CPU resources during periods of high CPU utilization. Second, we can prioritize events so that during high utilization, Services with higher priority get events routed to them in preference to lower priority services. Third, we need to support Authentication, Authorization and Roles for Services to prevent unauthorized triggering of Emergency Services.

The SLEE specification mandates that events belonging to a given Activity must be queued and delivered in first-come first-served order. The Activity priority extension to the SLEE is based on prioritizing activities or related streams of events. All the events fired on a given Activity are given the same priority value. The Event Router consumes events from multiple queues by selecting the event to be processed from the higher-priority non-empty queue as suggested in [draft-ietf-resource-priority-08]. To make the suggestion concrete we explain the operation in terms of SIP.

When a call is set up, the User Agent (IP Phone initiating the call) issues a SIP INVITE Message. The Call-ID header of the outgoing SIP INVITE identifies the call is associated with a SLEE Activity that is created for the call. A high-priority signaling message will have a Resource-Priority header. However, this is clearly subject to abuse by ordinary (non-emergency) users. Thus the SIP Resource Adaptor needs to authenticate and the SIP Message and authorize the user before believing the Resource-Priority value. After successful authentication, the event is assigned to the appropriate high priority queue according to the value of the header, otherwise the dropped as a spurious call setup attempt. After authentication all subsequent signaling messages related to the call may receive priority treatment. Note that these messages will all be identified with the same Activity and that the activity lasts for the duration of the call. (If mobility is supported at the Signaling layer, there could be several such messages as the emergency responder moves around in the affected area). As recommended in [draft-ietf-resource-priority-08] six different priority queues are supported in order to implements the "ETS" namespace. In addition there is one default priority queue for ordinary users.

During periods of high utilization, it may become necessary to pre-empt the execution of certain low-priority SBBs to allow higher priority SBBs more execution cycles. To do this, we devise mechanisms that limit the CPU utilization of an individual SBB by placing monitoring calls directly in the SBB code at the end of every basic block or non-branching sequence of bytecode instructions. We can accomplish this using bytecode re-writing at the time the SBB is installed.

Certain sensitive information such as location data and location of other emergency personnel may only be accessible by Authorized emergency personnel. These resources may implement their own authentication mechanisms at an application layer. However, we would like to support an overall policy framework that can be administered by the SLEE Administrator in a uniform way and without detailed knowledge of the authentication mechanisms of each of the resources accessed by the SBB. To support this we add the notion of Roles and Permissions to SLEE. Each SLEE service is assigned Permission. Roles and Permissions are comparable with each other. Each call is assigned a Role after the user is authenticated. The Service may only be instantiated if the Role of the user exceeds the Permission of the Service. Because the SLEE is protocol agnostic, we need to support this feature in such a way that our mechanisms will apply to a variety of Resource Adaptors and Services. For this we use the Java Authentication and Authorization Framework (JAAS). Details of our design are deferred to a more detailed paper.

6   Improving Quality of Service for Emergency Calls

The IETF document [4] describes call prioritization via the resource priority header and explains integration of resource management and SIP. A SIP User Agent (i.e. IP Phone or IM agent) must distinguish an emergency call and make arrangements for better quality of service for that call. We need an efficient and secure coupling between QoS supporting networks and the User Agent that originates the emergency call. As a minimal mechanism, we can simply drop the normal calls currently in progress to improve the chances of acquiring enough bandwidth for the high priority emergency call. Signaling protocols such as SIP support third party call control [5] which can be used to accomplish this. A BYE signal can be sent to both communicating parties resulting in the call being dropped. This is clearly a simplistic solution which users may find objectionable. If the underlying network supports a QOS mechanism such as DIFFSERV [6], a better solution may be devised.

With DIFFSERV, routers give preferential treatment to marked IP Packets. However, such a scheme is subject to abuse. Only authorized personnel should be allowed to originate IP Packets that get priority treatment at routers. To deal with this, we introduce an IP to IP media gateway as shown in the Figure 2.

The job of the media gateway is to mark packets originating from authenticated IP addresses that it knows about. The core network will only accept packets marked by the media gateway. The media gateway is in turn informed by the signaling server (i.e. service inside the SLEE) about authenticated and authorized users (i.e. IP Source addresses) from which it may accept incoming packets for marking. Non-emergency call packets are not marked and hence enter the core network as such and receive best effort service. The core network rejects any IP Packets that are marked for priority handling which do not originate from the gateway. The gateway thus functions as an admission controller – only allowing packets from authorized emergency personnel to be marked for better quality of service.

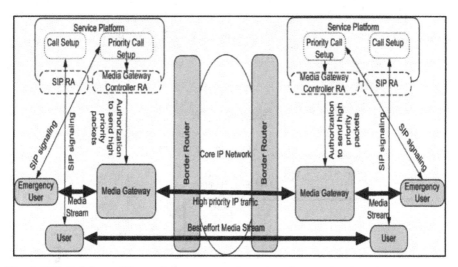

Fig. 2. ETS Call Setup. The media gateway marks packets from authenticated users. The Border gateway controller only admits marked packets from the media gateway

The interactions are as follows:

- The SIP Resource Adaptor of the SLEE authenticates the user and verifies that the user is an authorized emergency worker with a need for resource priority.
- The SLEE provides a privileged service that can only be invoked by authorized callers. When the call setup for the call comes in, the privileged service communicates with the QOS gateway and instructs it to mark packets originating from
- The IP address and port that are specified in the SDP portion of the SIP message.
- The media gateway trusts the SLEE and maintains a table of authorized IP addresses and ports and rejects any packets that do not originate from these addresses.
- The SLEE signals the Authenticated UA to send media to the QOS gateway rather than directly stream media to its peer.
- The media gateway marks any packets that originate from the Authenticated UA for priority treatment and rejects other packets.
- The core network routers only accept packets that are marked for priority treatment that originate from authorized media gateways.

The interactions are shown in more detail in the Figure 3. Scalability issues may be addressed by increasing the number of such media gateways. The signaling server picks the appropriate media gateway based on current load.

7 Conclusions

In this paper we proposed a standards-compliant solution for support of an IP Emergency Telecommunication system. We have limited our attention to end-to-end IP calls. Our proposal includes signaling prioritization and media prioritization without reserving resources a-priori at the network. We propose coupling call setup signaling with setting up improved quality of service for emergency workers. Our solution does

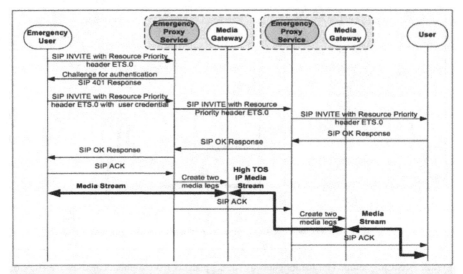

Fig. 3. Signaling interactions for ETS Call Setup

not consider congestion at access networks – i.e. the call setup would compete with non-privileged users. This may be addressed by dedicating an access network for use by emergency workers and is outside the scope of our proposal.

We are implementing a demonstration prototype of the mechanisms outlined in this paper in the Mobicents [7] open source SLEE platform. For the border gateway, we are using IP Tables to do QoS packet marking. The core network is emulated using NIST-NET network emulator.

Acknowledgement

The contributions of K. Sriram of the Advanced Networking Technologies Division are gratefully acknowledged.

References

1. GETS Government Emergency Telecommunications Support. http://gets.ncs.gov/
2. RFC3261 Session Initiation Protocol http://www.ietf.org/rfc/rfc3261.txt
3. Java Community Process JSR 22: The JAIN Service Logic Execution Environment.
   http://www.jcp.org/en/jsr/detail?id=22
4. Communication of Resource Priority for SIP
   http://www.ietf.org/internet-drafts/draft-ietf-sip-resource-priority-08.txt
5. RFC3725 Best practices for 3rd party call control in SIP.
   http://www.ietf.org/rfc/rfc3725.html
6. RFC2475 An architecture for Differentiated Services http://www.ietf.org/rfc/rfc2475.html
7. The Mobicents open SLEE platform, http://mobicents.dev.java.net/

On the Interaction of SIP and Admission Control: An Inter-domain Call Authorization Model for Internet Multimedia Applications

Ana Elisa Goulart and Randal T. Abler

Georgia Institute of Technology, Atlanta GA 30332, USA
agoular@yahoo.com, randal.abler@ece.gatech.edu

Abstract. In networks that support service differentiation, each in their own way, it is very important to manage how interactive multimedia applications use the networks' enhanced services. Therefore, ways to ensure that these services are properly authorized and accounted for are needed. This paper addresses the role of Session Initiation Protocol (SIP) proxies to authorize QoS-enabled multimedia sessions, based on the session's policy information and the network resources' availability. Our inter-domain call authorization model provides call authorization status and adds more granularity to the authorization process. This model is implemented in a SIP testbed, and simulation results showed that the model is scalable at end domains.

1 Introduction

In the SIP architecture, user agents (UAs) must be authenticated and authorized for every call or session request they make. Usually, SIP proxy servers perform the role of authentication and authorization [8]. The proxy's role of call authentication and authorization can be combined to the role of an *application-layer admission control*. This means that in the initial steps of the call setup process the application layer interacts with the network layer entities to verify in a higher level if the network can admit the call. Basically, what has just been described is very similar to the idea of *gate controllers*. The concept of gate controllers was introduced in the DOSA network architecture for IP telephony [4]. The *gates* are the ones that control and provide access to network resources, such as edge routers that act as policy enforcement points (PEPs) [10]. *Gate controllers* make the decision on whether the *gates* should be opened and set up the *gates* to admit the new call.

The architecture in which the gate controller concept was originally proposed is a generic signaling architecture and does not address the details of the SIP architecture. A more recent work by the IETF (RFC3313) [5] defines a SIP extension that can be used to integrate QoS admission control with call signaling. This extension consists of a new header field called *P - Media-Authorization* header, which the proxy attaches to a SIP message to inform the UA of the results of QoS media authorization (Figure 1). In RFC3313, the interaction between

T. Magedanz, E.R.M. Madeira, and P. Dini (Eds.): IPOM 2005, LNCS 3751, pp. 9–18, 2005.
© Springer-Verlag Berlin Heidelberg 2005

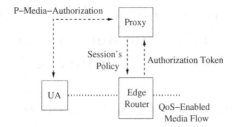

Fig. 1. Basic architecture for media authorization (RFC3313)

application layer and network layer is mainly in terms of transferring session's policy information and the results of call admission (as a media authorization token) on a preliminary policy-based admission control in the interface between SIP proxies and PEPs.

In the interaction of SIP and admission control, there are additional services that can be provided to enhance the media authorization model proposed in RFC3313. Here are some aspects that are considered in this paper:

- *Call authorization status at the destination domain:* In the scheme proposed in RFC3313 the media authorization token is transferred only inside a do- main, from proxy to user agent through the *P-Media-Authorization* header. Therefore, the destination domain has no information on the call authoriza- tion status at the origin domain. Since the signaling path is independent from the data path, potential problems may occur when the signaling path does not include the proxy that would interact with the network for media authorization.
- *Caller's account information:* Additional granularity to the call authoriza- tion process is desired in an open environment of untrusted domains and/or untrusted users (e.g., mobile users), according to [9]. Also, the destination domain requires additional information about the caller's identity in order to authorize the call completion or not.
- *Signaling impact on the call setup process:* The additional signaling needed in the interaction of SIP proxies and the network layer to obtain media authorization requires an analysis of its impact on the call setup process.

In this paper we address the need of integration of origin and destination do- mains in the call authorization process; thus, call authorization in each remote access network is part of a larger call authorization model: an *inter-domain call authorization model.* Targeting in this direction, an implementation of a call au- thorization model that combines the concept of gate controllers [4] to the media authorization model proposed in RFC3313 [5] is presented in this paper. In the proposed implementation of SIP proxies as gate controllers, the goals are to overcome the aforementioned issues of communicating call authorization status to destination domains and adding more granularity to the call authorization process. In addition, the model is tested in a SIP testbed, assuming a univer- sity network scenario that has remote and independent campus locations where

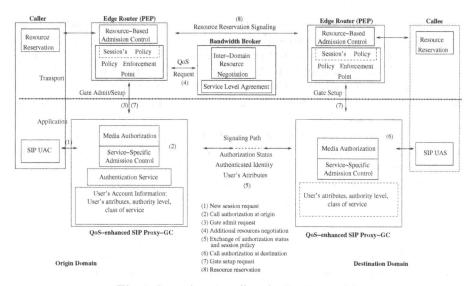

Fig. 2. Inter-domain call authorization model

multimedia sessions may be established between them and users may move temporarily from one location to another. This scenario assumes a set of common policies among end domains; however, there may be untrusted domains in the path between origin and destination domains. Scalability issues are also addressed in a simple queuing analysis where the impact of the additional service load on the SIP proxy servers is evaluated.

2 Inter-domain Call Authorization Model

To inform the destination domain that the call has been authorized (in special media authorization), an end-to-end approach to media authorization is to have a SIP proxy in each domain that strictly performs the role of a gate controller – a *QoS-enhanced SIP Proxy-GC* – and exchanges policy control information and call authorization status to trusted domains. An overview of the call authorization model here proposed is presented in Figure 2, which highlights the functionality of the proxy layers for call authorization. The steps needed in a call setup with inter-domain call authorization are also shown.

As shown in the block representing the QoS-enhanced SIP proxy-GC, the top of the structure includes service-specific admission control and media authorization. They both rely on an authentication service, which in turn relies on user's account information that is stored in a local database. This local information can be obtained during the user's registration process, where in addition to informing its current location, the user can inform its media capabilities and attributes for a more granular authentication process.

In the proposed implementation, policy information and call authorization status are transferred between domains so that end users and proxies work together in the role of inter-domain call authorization. Policy information (e.g., user's authority level) and call authorization status are carried in a new header *P-Auth-Profile* added by the SIP proxy and transported over the network to inform the destination domain if the media flows have been authorized at the origin domain, and vice-versa. The new header is used only in those requests and responses that can carry a SIP offer/answer (in a similar way as the *P-Media-Authorization* header [5]).

As of the current implementation, the *P-Auth-Profile* header carries two parameters: the user's authority which is here defined based on our university campus network scenario (e.g., STUDENT, FACULTY, STAFF), and the call authorization status that explains whether the request has been authorized in the source, destination, or both domains (e.g., SRC, DEST, SRCDEST).

2.1 Call Signaling Flow

Figure 3 illustrates the call signaling flow at the origin and destination domains. Following the succession of events, at the origin domain these are the functions the QoS-enhanced SIP proxy-GC performs to authorize a new call request:

- *User authentication (1):* After the initial authentication handshake [8], the proxy verifies the user's identity by querying a database that has the user's account information. This information may be limited only to the authority level and class of service of the user. For a new user in the domain, this information can be loaded during registration and its authenticity then verified with the user's home domain.
- *Service-specific admission control (2):* Based on the user's information, the type, and priority of the request, the proxy verifies the domain's local forwarding policies to verify, for instance, if this type of call can be serviced in this domain, if the number of calls the proxy can serve has not reached the maximum allowable limit, or if there are restrictions on forwarding the request to the destination.
- *Media authorization – Gate Admit Request (3):* The proxy interfaces with the network to verify if it can admit this call, in terms of required resources and the type of user. The proxy gives information to the network such as the bandwidth required for the call, the class of service, the call priority, and the authority level of the user. The network verifies the amount of aggregated resources already allocated for the class.

Then, the proxy forwards the request with additional policy control information in the *P-Auth-Profile* header. Upon receiving the request, the QoS-enhanced SIP proxy-GC at the destination domain performs the following functions:

- *Service specific admission control (4):* Being informed through the *P-Auth-Profile* header of the user's attributes and that the request has been admitted to the network at the origin domain, the proxy at the destination domain makes the decision on whether to forward the request to the final destination. Local policies may define a maximum number of users in a given multimedia session, or only users at a given authority can make this type of call.

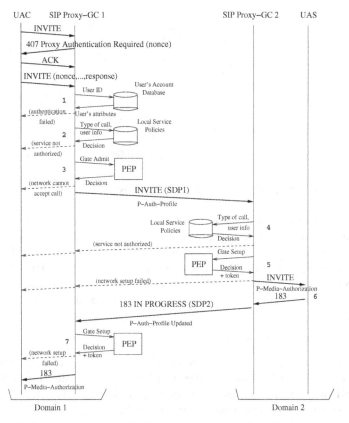

Fig. 3. Call signaling flow with the inter-domain call authorization model

– *Media Authorization – Gate Setup (5):* The proxy sends a gate setup request to the edge router with the policy information associated with the session information that arrived in the request. The edge router saves this information for use when it receives a resource reservation request for the flows associated with this call.

The proxy then forwards the request to the final recipient of the call request. This user then performs the media capability negotiation to decide which media flows it can communicate (6). Then it issues a response with a new session description (*183 Session In Progress* response, with SDP2 in the body of the message, and an updated *P-Auth-Profile* header). When this response arrives at the origin domain, the proxy sets up the gate in a similar way as it is done at the destination domain (step 5), with the updated information on the media capabilities and the call authorization status (7).

When the resource management phase begins, reservation requests are triggered at the user agents at both the origin and destination domains, as per RFC3312 [2]. The edge routers then verify if the media has been authorized and if its gates have been set up for the call.

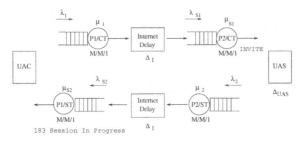

Fig. 4. Queuing model

3   Call Setup Delay Analysis and Scalability Issues

This section addresses the additional service time imposed on the proxies, the effects on SIP messages retransmissions, and scalability issues in terms of the number of additional proxy servers needed to meet the service load in the network. Moreover, this analysis is based on the assumption that UDP is the transport protocol, which means that a message-based transmission is assumed, and the reliability of SIP messages is handled at the application layer [7].

3.1   Queuing Analysis

Consider the queuing model illustrated in Figure 4. In this model, we assume that the processing of SIP messages takes considerable time due to queuing delays at the SIP agents and servers. This approach has been taken by Banerjee *et al.* [1] to study the handoff performance of SIP for handling mobility. Thus, we adopt the proposed queuing model to estimate the delays due to the queuing of SIP requests and responses as a consequence of the increased service load at the QoS-enhanced SIP proxies-GC at the end domains.

The queuing delays are computed based on the assumption that the proxy servers perform dedicated jobs (i.e., to process SIP messages) and that they can be modeled as an M/M/1 queue. The Internet delays are assumed to be constant. Also, the model assumes separate queues within the same proxy server in the processing of requests and responses.

The round-trip delay of the initial SIP *invite* request to the arrival of the *183* response $(T_{Inv-183})$, can be computed as

$$T_{Inv-183} = T_{P1/CT} + \Delta_I + T_{P2/CT} + \Delta_{UAS} + T_{P2/ST} + \Delta_I + T_{P1/ST}.$$

Based on M/M/1 queuing systems, in steady state, the average delay per message (waiting time in queue plus service time) is

$$T = \frac{\rho}{\lambda(1 - \rho)} \tag{1}$$

where $\rho = \lambda/\mu$ is the service load or utilization factor.

Now, consider a proxy server's processing time whose service rate (in messages/sec) is decreased from μ to $K\mu$, where $K \in (0, 1]$. It follows that the utilization factor ρ also decreases by a factor of K and the new queuing delay (T_{new}) increases by

$$\frac{T_{new}}{T} = \frac{1-\rho}{K-\rho}. \tag{2}$$

It is important to notice that the steady state equations being used assume that $\rho < 1$, i.e., $\lambda < \mu$. Therefore, if the new service rate μ decreases considerably, the utilization factor $\rho = \lambda/\mu$ tends to 1. When the arrival rate λ becomes bigger than the service rate μ, then a new server will be needed to avoid excessive queuing delays. In this case, the M/M/1 queue becomes an M/M/m queue, with utilization factor $\rho = \rho/m$.

3.2   Numerical Results

In the experiments, the QoS-enhanced SIP proxy-GC is tested in a SIP testbed [3], which has one user agent and proxy server in each end domain. The call setup procedure follows the signaling proposed in RFC3312 [2]. Also, we assume a reservation delay of 500 $msec$ (D_{res}) and a user delay to answer the call of 1 sec (D_{ans}).

Table 1 shows the results of the average overall call setup delay T_{SIP} and the round-trip delay of the initial *invite* request to the arrival of the *183* response $T_{Inv-183}$ with and without the implementation of the proposed inter-domain call authorization model. The delay increase with the implementation of the inter-domain call authorization model is mainly a result of the delay increase at proxies P1 and P2 in the initial signaling exchange of the SIP call setup transaction.

Table 1. Call setup delays measured in the testbed

	RFC3312	RFC3312 with proxies-GC
T_{SIP}	2132 $msec$	2358 $msec$
Ratio	1.00	1.11
$T_{Inv-183}$	128 $msec$	157 $msec$
Ratio	1.00	1.23

Based on the experimental results obtained through the testbed, the simulations of the queuing model in steady state initially adopt the increase delay ratio as 1.2 and get its inverse (0.833) as the factor K that decreases the service rate at the proxies. Then, different values of K will be considered, up to a 50% delay increase in the service time ($1/K = 1.5$). In addition, we use the following parameter values: service time $1/\mu = 20$ $msec$; Internet delay $\Delta_I = 150$ $msec$; and processing delay at user agent server (callee) $\Delta_{UAS} = 10$ $msec$. To simplify the analysis we assume that $\lambda_{S1} = \lambda_2 = \lambda_{S2} = \lambda$, and that the service rate at the proxy servers μ is affected by the factor K at queues P1/CT, P2/CT, and P1/ST.

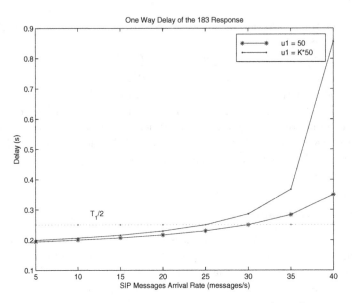

Fig. 5. One-way delay of the *183* response (μ=50)

In the exchange of the initial *invite* request and the receipt of the *183* response, there are time constraints on the receipt of the *183* response to avoid retransmissions: the UAS waits $T1$ seconds between the sending of the *183* response and the arrival of the provisional response acknowledgement message (PRACK) [6]. Hence, an approximation can be made that the one-way delay for the transmission of the *183* response between the UAS and UAC must be less than $T1/2$ (i.e., 250 *msec*, based on the standard's default value) to avoid any retransmissions. In Figure 5, the one-way delay from UAS to UAC of the *183* response is depicted for a service rate of μ increasing from 5 to 40 messages/sec, and for a reduced service rate $K\mu$ at proxy P1. First, the incoming rate of SIP messages must be limited to 30 requests/sec in order to avoid delays greater than $T1/2$. Then, in the case of reduced service rate at proxy P1, the maximum number of incoming call requests is reduced to 25 requests/sec.

Now, we consider additional servers at proxy P1. The results of maximum SIP messages arrival rate *vs.* the number of servers (m=1,2,3,4) are summarized in Figures 6 and 7. Figure 6 shows how the parameter K impacts the maximum SIP message arrival rate as new servers are added. For instance, when the increase in service rate is 50% ($1/K = 1.5$) the number of SIP messages that will be serviced within the delay constraints increases at an average rate of about 36 new messages per new server added. On the other hand, Figure 7 shows the effect of a higher service rate at the proxy for a fixed parameter $1/K = 1.2$. The higher service rate μ=100 that was tested increased the number of SIP messages arrival rate in comparison with μ=50.

The model imposes a higher processing load at the proxies, which may lead to the need to add more servers in the network to cope with an increasing

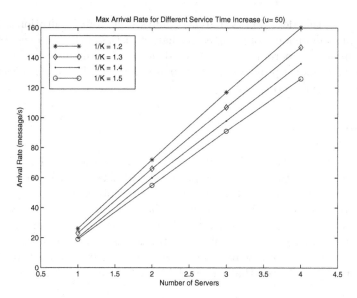

Fig. 6. SIP messages arrival rate for different values of service time increase (1/K)

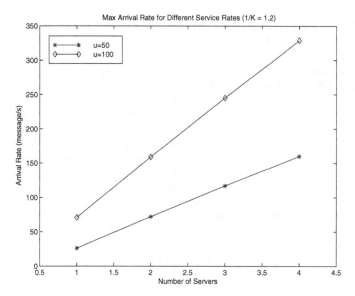

Fig. 7. SIP messages arrival rate for different values of service rate (μ)

number of incoming SIP messages to be processed. However, as the number of servers increases, the number of SIP messages that will be serviced within the delay constraints to avoid retransmissions increases linearly. These results lead us to the conclusion that the inter-domain call authorization model is a scalable model.

Of course, this delay and scalability analysis has some limitations, since the results can be dependent of the specific characteristics of the implemented tasks, and the network's parameter values used in the current model. However, the previous experiments provide a first estimate on the performance aspects of the proposed implementation.

4   Conclusion

This paper presented a new inter-domain call authorization model that explores the two-way interaction between application and network layers in the process of call admission control with QoS guarantees. This new model incorporates the concept of a gate controller [4] and media authorization [5]. Based on those two concepts, and adding features to ensure that the call authorization status is transferred between end domains and to increase the granularity of the authorization process, the new model has been implemented in a SIP testbed and simulation results showed that the model is scalable at end domains.

References

1. N. Banerjee, W. Wu, K. Basu, and S. K. Das. "Analysis of SIP-based Mobility Management in 4G Wireless Networks". *Computer Communications*, 27:697–707, 2004.
2. G. Camarillo, W. Marshall, and J. Rosenberg. "Integration of Resource Management and Session Initiation Protocol (SIP)". *IETF RFC 3312*, October 2002.
3. A. Goulart and R. T. Abler. "On Overlapping Resource Management and Call Setup Signaling: a New Signaling Approach for Internet Multimedia Applications". *Computer Communications*, 28(8):851–863, 2005.
4. P. Goyal, A. Greenberg, C. Kalmanek, W. Marshall, P. Mishra, D. Nortz, and K. Ramakrishnan. "Integration of Call Signaling and Resource Management for IP Telephony". *IEEE Network*, pages 24–32, May/June 1999.
5. W. Marshall. "Private Session Initiation Protocol (SIP) Extensions for Media Authorization". *IETF RFC 3313*, January 2003.
6. J. Rosenberg and H. Schulzrinne. "Reliability of Provisional Responses in the Session Initiation Protocol (SIP)". *IETF RFC 3262*, June 2002.
7. J. Rosenberg, H. Schulzrinne, G. Camarillo, A. Johnston, J. Peterson, R.Sparks, M. Handley, and E. Schooler. "SIP: Session Initiation Protocol". *IETF RFC 3261*, June 2002.
8. S. Salsano, L. Veltri, and D. Papalilo. "SIP Security Issues: The SIP Authentication Procedure and its Processing Load". *IEEE Network*, 16(6):38–44, 2002.
9. D. C. Sicker, A. Kulkarni, A. Chavali, and M. Fajandar. "A Federated Model for Secure Web-Based Videoconferencing". In *Proc. IEEE Conf. on Information Technology: Computers and Communications (ITCC'03)*, pages 396–400, 2003.
10. R. Yavatkar, D. Pendarakis, and R. Guerin. "A Framework for Policy-Based Admission Control". *IETF RFC 2753*, June 2002.

Experimental Evaluation
of the IP Multimedia Subsystem

Adetola Oredope[1], Antonio Liotta[1], Kun Yang[1], and Daniel H. Tyrode-Goilo[2]

[1] University of Essex, Department of Electronic Systems Engineering
Wivenhoe Park, Colchester, CO4 3SQ, UK
{aoredo,aliotta,kunyang}@essex.ac.uk
[2] University of Surrey, Guildford GU2 7XH, UK
dtyrode4@yahoo.co.uk

Abstract. The IP Multimedia Subsystem (IMS) is the latest framework for a seamless conversion of the ordinary Internet with mobile cellular systems. As such it has the backing of all major companies since it aims to offer a unified solution to integrated mobile services, including mechanisms for security, billing, quality of service and so forth. We provide a unique assessment of IMS based on our experimental test-bed, investigating functional and performance capabilities. Our study helps evaluating the level of maturity of state-of-the-art open source technologies in view of the deployment of IMS. We assess the suitability of SIP, IPv6, MIPv6 and IPsec as basic IMS enablers, highlighting crucial shortcomings which need immediate attention.

1 Introduction

The IP Multimedia Subsystems (IMS) is considered by both network and service operators as a platform to bring about the long awaited all-IP convergence of the cellular world and the Internet [1]. IMS carries both signaling and bearer traffic in which multimedia sessions can be created, modified or deleted delivering voice, data and multimedia contents to end users [2].

As the number of devices that connects to the Internet via Third Generation (3G) networks increases [3], IMS has the advantage of providing Quality of Service (QoS), better billing system and integration of services which a 3G *per se* cannot offer [1]. Given the substantial interest of IMS, we have carried out a pilot study aimed at assessing its core functionality. This was done by building a test-bed based on the specification of IMS as described by the Third Generation Partnership Project (3GPP) Release 5 [4] which mandates the Internet Protocol version 6 (IPv6) for increased addressing space, Mobile IPv6 (MIPv6) for mobility, Internet Protocol Security (IP-Sec) for security and Session Initiation Protocol (SIP) for signaling and bearer traffic. We have integrated these as an overlay network allowing transparent connectivity between fixed and mobile networks.

Previous work has studied the integration of these components but not in light of IMS. Examples include the integration of MIPv4 with security [5], MIPv6 with SIP [6], IPv6 with SIP [7], and mobility with SIP [6]. Some limitations have been revealed such as the communication problems introduced by Network Address Translation (NAT), the interference of foreign agents [5], handoff [6], user agent registration [7] and handover [8].

T. Magedanz, E.R.M. Madeira, and P. Dini (Eds.): IPOM 2005, LNCS 3751, pp. 19–28, 2005.
© Springer-Verlag Berlin Heidelberg 2005

This fact motivated us to integrate the aforementioned protocols together on a single platform over a multi-access network (LAN, WLAN, UMTS and GPRS) to emulate a typical IMS environment. Experiments were then carried out to determine the level of maturity of the individual components, the interoperability of the components and how the platform would respond to issues like registration, mobility and handover in different environmental scenarios like peer-to-peer file transfer, client-server services and voice services. This revealed a number of limitations in the IMS architecture such as the breaking of certain applications in pure IPv6 native modes, failure of bidirectional tunneling, and the flushing out of routing tables immediately after handover.

Section 2 gives a brief description of background IMS technologies for the benefit of readers who are not familiar with them. Section 3 describes our test-bed, experimental setup and results. Conclusion and recommendations are drawn in Section 4.

2 Background Technologies

2.1 Internet Protocol Version 6 (IPv6)

The Internet Protocol version 6 (IPv6) [9] was designed to address most the limitations found in IPv4 [10], mainly its limited addressing space (IPv6 uses 128 bits), simple and self routing headers, extension headers for optional services, quality of service and security [9]. IPv6 has been widely deployed commercially by most operators [7]. Many standard bodies are considering IPv6 for their next generation networks and services. For example 3GPP mandates IPv6 for IP Multimedia Subsystem (IMS) in Release 5 [4]. Operating systems (Microsoft Windows [21] and Linux [10]), routers (Cisco [22]), and other network elements are starting to support IPv6, hence our choice to adopt IPv6 in our IMS test-bed.

2.2 Mobile IPv6 (MIPv6)

One of the most important challenges IP networks face is mobility support. Mobile IPv4 (MIPv4) [11] and Mobile IPv6 (MIPv6) [12] are the results of IETF mobility support efforts. Mobility protocols enhance the network layer, so that IP hosts can change location and retain their communicating sessions. But MIPv4, which is built on IPv4, has certain limitations in terms of performance, scalability and reliability. The packets for the Mobile Node (MN) need to be tunneled via the Home Agent (HA) when the MN has left its home network which introduces a bottleneck and a single point of failure, in addition to being inefficient (triangle routing) [12]. There are also certain issues of packets being dropped by ingress filters in foreign networks because the MN's address is not recognized and also the Foreign Agent inability to process/forward the MN's registration packets to it's HA, forcing packets to be dropped [8]. These issues are addressed in MIPv6, where packets are allowed to register directly with Corresponding Nodes (CN) and triangular routes can be avoided (Route Optimization) through routing headers which lead to performance degradation [12]. For our purposes, MIPv6 was selected just for its support of terminal mobility, since current MIPv6 implementations do not have any security mechanisms in place to protect the payload [13].

2.3 Internet Protocol Security (IPSec)

Given the specifications of IMS we have selected IPSec [4] among other security protocols [14]. Also IPSec is the preferred choice for IPv6 [9] and MIPv6 [12]. It is an extension of IP [15] which allows security to be removed from the network and placed on the end points [14] by applying encryption to either the entire IP payload (*tunnel mode*) or only the upper-layer protocols of the IP payload (*transport mode*). IPSec allows for security services such as data integrity protection, data origin authentication, anti-replay protection and confidentiality, offering also protection to the upper layers [15].

IPSec uses two protocols. The Encapsulation Security Protocol (ESP) provides confidentiality, data integrity and data source authentication of IP packets by encapsulating the data to be protected. The Authentication Header (AH) protocol supports data integrity and data source authentication but does not offer any form of confidentiality, which makes it a lot simpler than ESP.

2.4 Session Initiation Protocol (SIP)

Most Internet applications involving multimedia elements such as voice, video and data are built on the Session Initiation Protocol (SIP) [16], proposed by the IETF as the protocol for handling call setup, routing, authentication and other feature messages to endpoints within an IP domain [1]. Applications using SIP include Voice over IP (VoIP), distributed games, and white boarding among others [2].

SIP is modeled on HTTP and uses ASCII text messages. It contains headers supporting MIME to allow it to work with exiting Internet applications [16]. A session is established via a three-way handshake. A TCP connection is first created by the caller to its recipient; then an INVITE message is sent to the other client. The message contains the caller's capabilities, media types and formats. The other client replies with a message including its own capabilities, media types and format. SIP has now reached a very high level of maturity and is widely adopted on most IP based applications. 3GPP mandates its use for multimedia session negotiation and session management [4].

2.5 IP Multimedia Subsystems (IMS)

The IP Multimedia Subsystem (IMS) is an all-IP overlay network with the aim of converging fixed and mobile networks into a single access independent network [1]. IMS easily achieves this goal because it is built on SIP for session management and control. SIP allows clients to INVITE other clients into a session and it then negotiates control information, terminal capabilities and media channels to be used in the session. Then IMS allows the clients to be connected via IP. The main functions of IMS are to allow for authentication and authorization of both fixed and mobile clients, support roaming, charging, and Quality of Service (QoS), providing access and network domain security [17]. IMS facilitates the deployments of applications such as Voice over IP (VoIP), Instant Messaging, Push-to-talk, Video Conferencing, Presence and Online Gaming.

3 Evaluation

3.1 Test-Bed Setup

Our test-bed was implemented on Linux kernel 2.4.22 based on readily available open source technologies. The *UniverSAl playGround for Ipv6* (USAGI) [18] release 5 was used for the implementation of IPv6 and IPSec because of its support for multiple paths to the same destination with equal or unequal metrics and its simplified method for adding and removing default routes which is not supported in other implementations. The IPSec implementation allows for the ESP and AH protocols to be used. The AH mode was not tested because it has lost commercial value since it does not provide confidentiality. The ESP-transport mode was used for the binding updates, registration and other forms of signaling and the ESP-tunnel mode was used for the data packets. Internet Key Exchange (IKE) protocol was used to negotiate and setup IPSec security policies so the shared keys could be generated dynamically and securely. We decided not to use USAGI's MIPv6 implementation because this is outdated and there are no plans for it to be updated [18]. Instead we implemented the Mobile IPv6 for Linux (MIPL) stable Release 1.1 [13] which contains a specific module for the HA, MN and CN. The CN features are implicitly contained by the HA and the MN.

SIP was implemented using SIPset from the Vovida SIP stack [19]. SIPSet is a SIP User Agent with a GUI front end that works with most SIP Servers and can be used as a soft phone, to make and receive VoIP phone calls [19].

To fully experiment with mobility over a multi-access network all network terminals requiring mobility (MN, HA and CN) were connected via WLAN 802.11b (5.5Mbps) and 100Mbps Ethernet LAN was used for the fixed networks. We also included an impairment node whose role was to emulate GPRS and UMTS conditions that are typically experienced in real networks.

3.2 Test 1: Platform Validation

Aim
The purpose of the first series of tests was to determine and verify the functionality, robustness and stability of the integration of IPv6, MIPv6, IPSec and SIP as an IMS Platform.

Setup
The Mobile Node (MN) and the Corresponding Node (CN) were both configured as SIP clients and IPSec was enabled. A call was then initiated between the clients. 22 different setup configurations were tested including IPv6/IPSec stateful and stateless address allocation in both transport and tunnel modes.

Results
Peer-to-peer calls were made successfully in all scenarios confirming the functionality of our IMS test-bed and it was possible to gather information like terminal capabilities and media channels. We have, however, evidence that our platform is stable and that it performs all basic IMS functions correctly.

3.3 Test 2: Horizontal Handover Tests

Aim

Validate and assess the functionality, stability and horizontal handover properties of our IMS test-bed in a WLAN environment.

Setup

A web server running Apache version 2 was configured on the Corresponding Node (CN) in order to perform downloads between end points. MIPv6 was then enabled on both the Home Network (HN) and the Foreign Network (FN) with routes to the web server as shown in the network setup depicted in Fig 1.

The test starts by initiating a file transfer between the MN – sitting in the home network – and the web server. During the transfer horizontal handover is forced. To emulate this process, the transition between Home and Foreign Networks was performed by setting different *Service Set Identifiers* (or SSID, a token used to identify an 802.11 WLAN network) and forcing the MN to change from one SSID to another. This procedure is known as *Hard Handover* since the IP point of attachment is terminated before the MN is moved to a new point of attachment (*break-before-make*). Using *ethereal*[1] we could identify the key MIPv6 packet exchanges during handover, determining their precise position in the file-transfer timeline.

A total of 22 different setup configurations were tested, assessing features such as MIPv6/IPSec in transport and tunnel modes using both manually configured keys (M.C.) and IKE. The abbreviations used below are summarized in Table 1.

Results

The total horizontal handover times w.r.t. pure MIPv6 with RO were comprised between 4.25 and 8.84 seconds with an average of 6.96 sec. Their distribution among individual factors is depicted in Fig.2 (legend of the x-axis in table 2). In the case of BT we obtained slightly better results (min=1.78 sec; max=8.47 sec; avg=4.68 sec) with a comparable distribution.

We can draw the conclusion that, the predominant factor in horizontal handover is the time taken by the MN to seize the router advertisement (RA) from the visited/foreign network to determine its prefix and the default router.

Table 1. Abbrevations

Abbreviation	Meaning
BU (MN to HA)	Binding Update
BA (HA to MN)	Binding Ackn
CoTI (MN to CN)	Care of Test Init
CoT (CN to MN)	Care of Test
BU (MN to CN)	Binding Update
Next-TCP-Packet	
RA	Route Advertisement
RO	Route Optimization
BT	Bidirectional Tunneling

Table 2. Key to Graphs

Key	Meaning
1	Pure MIPv6 RO Mode
2	Pure MIPv6 BT Mode
3	MIPv6 RO mode with IPSec M.C. transport Mode
4	MIPv6 RO mode with IPSec IKE transport mode
5	MIPv6 BT mode with IPSec M.C. transport mode
6	MIPv6 BT mode with IPSec IKE transport mode
7	MIPv6 RO mode with IPSec M.C. tunnel mode
8	MIPv6 RO mode with IPSec IKE tunnel mode

---

[1] http://www.ethereal.com/

BT incurs less overhead than RO in terms of MIPv6 traffic needed to restart home running TCP conversations (no CoTI/CoT required). In fact, in RO mode, the CoTI can become an issue of concern (if the application that is trying to reach the MN does not send packets during or shortly after handover, it may take considerable time for the MIPL implementation to send a CoTI).

Another determining factor is represented by the BA coming from the HA to the MN which can considerably increase when network performance degrades.

Looking at MIPv6/IPsec transport mode, the aggregate times (considering 9 different configurations not reported here for brevity) are: min=1.70sec; max=9.58sec; avg=5.47sec as shown in Fig 2, whereby the symbols in the x-axis are all explained in Table 2.

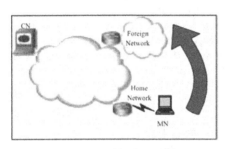
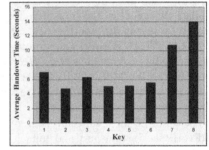

Fig. 1. Network Setup for WLAN handover   **Fig. 2.** Average time values for horizontal (WLAN/WLAN) handover

These highlight some issues. In the case of transport mode, in RO mode, CoTI takes a slightly higher percentage of the total handover time in comparison with pure MIPv6 BT mode. This is due to the processing overheads introduced by the IPsec. Because of this relative increase, the overall impact of waiting for a RA notification is reduced (8.5% on RO and 3.5% in BT) when compared with pure MIPv6. A similar negative impact of IPsec processing overheads is also experienced in BT mode. However, the general performance in this case is 5.9% better than RO.

Overall, when compared to pure MIPv6, RO with IPsec performed 19.1% faster, whilst BT with IPsec increased its average handover time by 11.8%.

In MIPv6/IPsec in tunnel mode, the *next-TCP-packet* value has considerably increased (from almost null to 66%). This can be explained by looking at how handover is implemented. During handover, routing tables are flushed and to regain connectivity after handover the private routing entries must be re-instated.

3.4 Test 3: Vertical Handover Tests

Aim 1

Validate and assess the functionality, stability and vertical handover properties between a WLAN and GPRS environment in the context of IMS.

Setup

The experimental setup was similar to the previous but includes an additional impairment node which emulates network conditions typically experienced in a GPRS network. The conditions relating to the following results are: Upstream Bandwidth =

10kbps; Downstream Bandwidth = 40kbps; Round Trip Time = 700msec. The handover takes place between WLAN (Home network) and GPRS (Foreign Network),

We implemented the impairment node based on a modular router for Linux platform, the Click Router Project [20]. Bandwidth metrics were verified with the *iperf* [2], a tool supporting IPv6 and capable of determining TCP/UDP bandwidth between two endpoints. The experiments were performed over the same (21) set-up configuration of Test 2.

Result

A representative sample of results is illustrated in Fig 3. The total vertical handover times w.r.t. pure MIPv6 with RO were comprised between 9.85 and 11.91 sec with an average of 11.08 sec.

In the case of BT we obtained slightly better results (min=5.9 sec; max=10.93 sec; avg=9.32sec) with a comparable distribution. In comparison to the corresponding figures obtained in the case of horizontal handover over WLAN (Sect. 3.3), the overall handover time is increased by 159.1% and 198.9% for RO and BT, respectively.

Once again, BT performed better (15.9%) on the total handover time. This performance degradation was expected to some extent, because GPRS conditions have an increased round-trip-time (RTT) (700msec against the few msec of WLAN). Also its bandwidth is significantly reduced (e.g. from 5.5Mbps to 40kbps, on the downstream); so any inward/outward communication to/from the MN will take significantly longer. The CoT and *next-TCP-packet*, increased in average from almost null values to 9% and 19% respectively. This reduces the percentage of impact that other parameters may have on handover, like RA, CoTI and BA which decreased by 20%, 5% and 3%, respectively.

Vertical handover in transport mode resulted in RO with {min=4.65 sec; max=12.05 sec; avg=8.77 sec} and BT with {min=4.00 sec; max=9.91 sec; avg=7.56 sec}. In comparison to the corresponding figures obtained in the case of horizontal handover over WLAN (Sect. 3.3), the overall handover time is increased by 55.7% and 42.2% for RO and BT, respectively. The CoT and *next-TCP-packet* parameters have increased to about 12%. In BT mode, however, the impact of the different MIPv6 elements on the overall handover showed no significant changes (below 5% in all cases).

Vertical handover in tunnel mode resulted in RO with {min=7.89 sec; max=19.36 sec; avg=16.21 sec}. In comparison to the corresponding figures obtained in the case of horizontal handover over WLAN (Sect. 3.3), the overall handover time is increased by 20.7%. The only difference is on the CoT which increased, on average form virtually 0% to 8.5%. This reduced the impact of the next-TCP-packet from 61.5 to 45%.

Aim 2

Validate and assess the functionality, stability and vertical handover properties between a WLAN and a UMTS environment in the context of IMS.

Setup

The experimental setup is identical to the previous one involving GPRS, with the only difference being the network conditions that were set to typical UMTS values: Upstream/Downstream Bandwidth = 128kbps; Round Trip Time = 500msec.

---

[2] Iperf: http://dast.nlanr.net/Projects/Iperf/

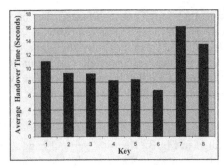
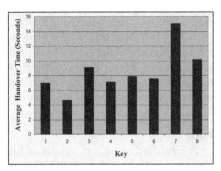

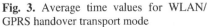

Fig. 3. Average time values for WLAN/ GPRS handover transport mode

Fig. 4. Average time values for WLAN/ UMTS handover transport mode

Results

A representative sample of results is illustrated in Fig 4. The total vertical handover times w.r.t. pure MIPv6 with RO were comprised between 2.93 and 11.39 sec with an average of 6.97 sec. In the case of BT we obtained {min=4.15 sec; max=4.92 sec; avg=4.64sec}. The distribution of overheads is similar to those of GPRS (differences below 5%). The UMTS figures are comparable to the case of pure WLAN (Sect.3.3) (less than 1% difference) but significantly better than GPRS. Vertical handover in GPRS is on average 158.1% slower in RO and 200.5% slower in BT.

WLAN/UMTS Vertical handover in transport mode resulted in RO with {min=5.07 sec; max=10.21 sec; avg=8.8 sec} and BT with {min=4.4 sec; max=10.18 sec; avg=7.7 sec}. The apparent slight performance degradation of UMTS was unexpected but can be entirely attributed to the WLAN. UMTS experiments were performed several months after their GPRS counterpart so it was difficult to obtain exactly the same conditions, given that we were using a WLAN shared by other users.

Our study highlights, however, a strong similarity between the distribution of overheads of UMTS and WLAN. A closer look at the UMTS results indicates that MIPv6 elements traversing the network are the predominant factor. Besides RA, the average time to trigger CoTI is also critical (these two affect 60% of the handover time).

Finally, vertical handover in tunnel mode resulted in RO with {min=7.49 sec; max=18.9 sec; avg=12.6 sec}. This places vertical UMTS handover between vertical GPRS handover and horizontal WLAN handover. This was expected, since the MIPv6 control packets incurred by the MN are beneficially affected by the better condition of UMTS. Our measurements help quantifying those differences. On average the proportion of next-TCP-packet is 49% (UMTS), 45% (GPRS) and 61.5% (WLAN). The overall handover time in UMTS in tunnel mode is on average 84.4% more efficient than GPRS.

4 Conclusion

The integration of the key IMS components and the experiments performed on the test-bed have enabled us to assess the level of maturity of IMS as a whole in relation to state-of-the-art open source technologies.

We have successfully integrated SIP, IPv6, MIPv6 and IPSec to work as an overlay network in a multi-access network environment. Even though there were some limitations (such as the flushing out of route tables by MIPv6 immediately after handover; the failure of IPSec to protect in certain scenarios; and the interference of source addresses during tunneling) these are actually functional problems which can be overcome by working on the individual open source technologies.

From the results obtained from our test-bed, IMS seems to be functional and set to be deployed commercially with the available open source technologies. SIP proved productive in obtaining the necessary parameters to be used by IMS while the integration of IPv6, MIPv6 and IPSec enables true secured mobility allowing for handover between access networks.

The biggest shortcomings are, instead, performance related. These arise from the fact that MIPv6 expects handover between networks to trigger with layer-3 awareness of change. This has significant (negative) impact on time-sensitive applications, which suggests that there is scope for improvement by looking at ways to trigger handover prior to the interruption of layer-3 communication. Relevant efforts are already pursuing this direction but, at the time of writing, there are no publicly-available solutions. The handover figures presented in this article derive from an MIPL implementation.

We plan to carry out further work in the implementation of QoS and charging with our test-bed as this will provide us with a platform to perform experiments on several IMS Services.

IMS as a whole has great potential in terms of bringing about the all-IP convergence anticipated by merging the internet to the cellular world and the available open source technologies available are quite mature for its deployment commercially.

Acknowledgments

The equipment used to build the test-bed has been provided by Vodafone Group R&D, U.K who has also suggested the figures on typical GPRS/UMTS network conditions. Particular thanks go to N. Papadoglou, H. Zisimopoulos, and O. Gurleyen (all from Vodafone) who have provided feedback and suggestions.

References

1. Camarillo, G., Garcia-Martin, M.A., "The 3G IP Multimedia Subsystem (IMS): Merging the Internet and the Cellular Worlds", Wiley, 2004.
2. "Siemens IP Multimedia Subsystem (IMS): The Domain of Services", Tech report, Siemens AG Germany, 2004.
3. Mikka, P., et al. "The IMS: IP Multimedia Concepts and Services in the Mobile Domain." Wiley, 2004.
4. "Shaping the future of mobile communication standards", 3GPP report, August – 2004.
5. Lauri, H., "Integrating Mobile IPv4 and IPSec authentication". Tech. Report, Helsinki University of Technology, April 2004.
6. Nakajima, N., et al., "Handoff delay analysis and measurement for SIP based mobility in IPv6". Proc. of IEEE ICC '03. 11-15 May 2003.
7. Flykt, P.; Alakoski, T. "SIP services and interworking with IPv6" 3G Mobile Communication Technologies, 2001. Second International Conference on (Conf. Publ. No. 477) 26–28 March 2001 Page(s):186–190

8. Jhonson, D., et al, "Mobility Support in IPv6" IETF RFC 3775, pp.1-39, December 1998.
9. Deering, S., et al., "RFC 2460 - Internet Protocol, Version 6 (IPv6) Specification". December 1998.
10. Bieringer, P., "Linux IPv6 HOWTO, a guide how to configure and use IPv6 on Linux Systems". GNU GPL version 2, pp. 1–119, March – 2004.
11. Adrangi, F., et al., "Problem Statement: Mobile ipv4 Traversal of VPN Gateways". IETF draft-ietf-mip4-vpn-problem-statement-03.i.txt, pp. 1–19, May 2004.
12. Perkins, C., "IP Mobility Support in IPv6". IETF RFC 3344,
13. "Mobile ipv6 for Linux", May 2004. www.mobile-ipv6.org
14. Ferguson, P., et al., "What is a VPN". The Internet Journal, Volume 1, Issue 1, June 1998.
15. Thayer R., et al., "RFC 2411 - IP Security", Nov. 1998.
16. Rosenberg, J., et al., "RFC 3261 - SIP: Session Initiation Protocol". June 2002.
17. "IP Multimedia – a new era in communication". Tech. Report, Nokia, 2004.
18. "Linux IPv6 Development Project", Technical Report USAGI Project, July 2004. www.linux-ipv6.org
19. "SIPSet", Tech. Report, Vovida, www.vovida.org
20. "The Click Modular Router Project". Technical Report, July 2004. http://www.pdos.lcs.mit.edu/click/
21. "Understanding Mobile IPv6" Microsoft Windows Server 2003, November 2004
22. "Technologies White Paper" Cisco, May 2001.

Open Service Access for QoS Control in Next Generation Networks – Improving the OSA/Parlay Connectivity Manager

Samson Lee[1], John Leaney[1], Tim O'Neill[1], and Mark Hunter[2]

[1] Institute for Information and Communication Technologies,
University of Technology, Sydney,
PO Box 123, Broadway NSW 2007, Australia
{samlee,jrleaney,toneill}@eng.uts.edu.au,
[2] Alcatel Australia
mark.hunter@alcatel.com.au

Abstract. The need for providing applications with practical, manageable access to feature-rich capabilities of telecommunications networks has resulted in standardization of the OSA/Parlay APIs and more recently the Parlay X Web Services. Connectivity Manager is an existing, 'stable' API for establishing QoS parameters in a network. However, it falls short as an interface to the expected requirements based on recent drafts of ETSI TISPAN's Resource Admission Control Subsystem. We analyze these requirements and suggest some improvements, which are incorporated into a new Parlay X Web Service specification. Furthermore, we describe our efforts to date in implementing a prototype of the specifications as well as our experience in utilizing the prototype to develop an example QoS-aware multimedia application.

1 Introduction

The traditional telecommunications environment is 'closed', where applications can only be developed internally with specific knowledge of individual network technologies. In the last several years, there has been an enormous increase in efforts to 'open up' these networks for application development [1]. In opening up the network, new business models emerge where applications can be developed and provided by enterprises outside the traditional network operator domain. These applications can utilize the feature-rich service capabilities of the network through standardized Application Programming Interfaces (APIs) with off-the-shelf IT technology and tools such as Java and Web Services. In this new environment, innovative new applications will reach the market with drastically reduced development cycles.

The Parlay APIs, otherwise known as Open Service Access (OSA), is a set of standardized open APIs that allow applications access to network functionality by packaging and presenting the service capabilities in a manageable fashion. The OSA/Parlay APIs are jointly developed and published by the Parlay Group

T. Magedanz, E.R.M. Madeira, and P. Dini (Eds.): IPOM 2005, LNCS 3751, pp. 29–38, 2005.

[2], Third Generation Partnership Program (3GPP) [3] and European Telecommunications Standards Institute (ETSI) [4], and form the API layer of the 3GPP IP Multimedia Subsystem (IMS). The APIs provide a technology agnostic abstraction of functions including call control, location and user interaction among others. Parlay X also achieves these goals, but further stimulates the development of next generation applications by IT developers who are not necessarily experts in telecommunications. Parlay X offers a higher level of abstraction compared to OSA/Parlay APIs, and exposes the interfaces through Web Services technology.

There are currently more than 208 announced products [5] implementing the OSA/Parlay APIs and/or Parlay X Web Services, including 26 gateways and 76 value-added applications, with support from a growing number of over 65 member organizations who are vendors, service providers and independent software developers. Clearly, these APIs are gaining wide acceptance in industry. However, current uses in both fixed [6, 7] and mobile [8] networks are focused on telephony-type applications, particularly those that require call control capabilities. Next Generation Networks will have advanced service capabilities such as QoS with admission control, multicasting and VPN provisioning. Next generation applications such as Video on Demand (VoD) and bandwidth on demand will require access to these advanced service capabilities.

The OSA/Parlay Connectivity Manager is an existing API for establishing QoS parameters in a network. Although it has been classified as a 'stable' specification for a number of years, it is still not widely supported compared to the other APIs. In spite of this lack of implementation, there is general agreement [9] by the Parlay Joint Working Group (consisting of members from ETSI, 3GPP and Parlay) about the need to continue research on an API for application-controlled QoS. Previously [10], we identified some major shortcomings of the Connectivity Manager as an interface to the expected requirements based on recent drafts from ETSI TISPAN's Resource Admission Control Subsystem, and suggested some improvements. This paper extends on our previous work by incorporating these improvements into a new Parlay X Web Services specification.

The remainder of this paper is organized in five sections. In Section 2, we provide an overview of the OSA/Parlay Connectivity Manager and our suggestions on improving it. In Section 3, we detail our new Parlay X Web Service for QoS control including our current efforts in implementing a prototype scenario and an example QoS-aware multimedia application. Finally, we summarize the paper and discuss future plans to evaluate and measure the effectiveness of the prototype and application development in Section 4.

2   OSA/Parlay Connectivity Manager

2.1   Overview of Current Specification

The OSA/Parlay APIs consist of the Framework and a number of Service Capability Features (SCFs). The Framework is the initial point of contact for the application and provides functions to discover and access the SCFs that are offered by the network. It also considers security precautions such as authenticating and authorizing applications. The SCFs are collections of interfaces that provide the application with access to the capabilities within the network. Connectivity Manager is one of the SCFs that have been standardized as a part of the OSA/Parlay specifications.

The OSA/Parlay Connectivity Manager [11] includes APIs between an *enterprise operator* and a *provider network* for the two parties to control QoS parameters. The specific mechanism to be used by the network provider is not mandated, although the specification mentions the use of Differentiated Services Code Point (DSCP) to identify aggregated traffic flows.

The network provider maintains a list of the enterprise operator's valid sites and Service Access Points (SAP) for which traffic flows through the provider network. The enterprise operator is able to establish traffic flows with specified QoS parameters between its SAPs. These flows are called Virtual Provisioned Pipes (VPrP), and the group of VPrPs is called a Virtual Provisioned Network (VPrN). Applications in the enterprise operator network may use the API to establish VPrPs based on pre-defined QoS templates. For instance, the provider may offer templates for video conferencing, high definition television, gold service, silver service, etc.

The top level `IpConnectivityManager` interface provides operations to get `IpQoSMenu` and `IpEnterpriseNetwork` menu interfaces.

The enterprise operator uses the `IpQoSMenu` interface to browse and configure pre-defined templates that describe the QoS parameters. The parameters in each template consists of the following datatypes: `TpPipeQoSInfo`, which defines the pipe's directionality, service origin, service destination, forward load and reverse load; `TpProvisionedQoSInfo`, which defines the delay, loss, jitter and excess load action; and `TpValidityInfo`, which defines the time, duration, days of week, and months for which the template is valid. The templates are negotiated in an off-line process, and values can be specified by either the network provider or enterprise operator.

The enterprise operator uses the `IpEnterpriseNetwork` menu interface to browse the list of its sites and SAPs (maintained by the network provider), and to manage its VPrN. Operations are provided to list, get details of, and delete current VPrPs, as well as to request the creation of new VPrPs. The status of VPrPs can be active, pending or disallowed. Details of each active VPrP contains the currently provided QoS parameters, which may be different to requested. Details of each pending or disallowed VPrP contains the requested QoS parameters. QoS parameters are defined by the datatypes described previously.

2.2   Suitability for ETSI TISPAN RACS Requirements

TISPAN (Telecoms & Internet Converged Services & Protocols for Advanced Networks) is the ETSI technical committee responsible for standardizing Next Generation Networks (NGNs). As a part of their work, ETSI TISPAN has proposed the Resource Admission Control Subsystem (RACS). RACS is the NGN subsystem responsible for elements of policy control, resource reservation and admission control. RACS also includes support for Network Address Translator (NAT) and Firewall (FW) traversal.

In our previous paper [10], we performed a detailed gap analysis to determine the suitability of using OSA/Parlay Connectivity Manager as an interface to satisfy ETSI TISPAN RACS draft requirements. The requirements are expected to evolve as work is progressed, but they should still provide an early indication of what to expect. The following summarizes our analysis:

1. **Control of resources based on access network capabilities.** Connectivity Manager satisfies this requirement by presenting an abstraction of the network capabilities to applications. It is possible to translate these abstracted functions into specific mechanisms implemented by the access network.

2. **Application Function (AF) reservation mechanism.** Connectivity Manager is able to provide AF with a mechanism to reserve resources through the open API. The open API would be logically located at a layer above the Gq interface.

3. **AF in multiple administrative domains.** The OSA/Parlay Framework provides the necessary management functions for applications located in multiple administrative domains.

4. **AF authentication and authorization.** Applications may need to be authenticated and authorized by the Framework before gaining access to the Connectivity Manager.

5. **Requests for traffic with specified directionality, symmetry and multicasting.** Existing OSA/Parlay Connectivity Manager does not support QoS requests for multicast traffic.

6. **AF notification of resource change.** Existing OSA/Parlay Connectivity Manager does not support notification of resource changes. Instead, application can only poll the server for network changes.

7. **AF modification of existing reservations.** Existing OSA/Parlay Connectivity Manager does not support modification of VPrPs.

8. **Admission feedback for AF.** Existing OSA/Parlay Connectivity Manager is limited to synchronous feedback messages, i.e. results are returned immediately. If the immediate result is 'pending', then the application will not be notified when a final admission decision is made, and will need to poll in order to obtain an updated status.

9. **Export charging information.** Charging requirement could be satisfied through OSA/Parlay Charging SCF. Modifications to the existing Charging SCF may be required.

10. **Push and pull control mechanisms.** QoS requests could be performed either by an application server or directly by the end-user.

11. **Session aware.** This means the RACS is able to interact with the Call Session Control Function (CSCF). This could be satisfied at the application level by using the Call Control SCF in conjunction with the proposed open API for QoS control.

12. **Dependency on network for policy enforcement.** Enforcement is done by the network. There is a dependency on the QoS functions working properly in the underlying network. We are publishing our work on Policy Based Network Management in separate papers, including [12].
13. **Priorities for multiple/conflicting requests.** Existing OSA/Parlay Connectivity Manager is limited to priorities for values inside `TpProvisioned Info` datatype only. There is no priority field inside `TpPipeQoSInfo` and `TpValidifyInfo` datatypes.

2.3   Suggestions to Improve the Specification

Based on the items listed in 2.2, Connectivity Manager is unable to meet RACS requirements because there is limited or no support for: modification of existing VPrPs (7); notification of network changes (6); asynchronous feedback of admission control messages (8); multicasting requests (5); and priorities for multiple/conflicting requests (13).

In order for Connectivity Manager to be more inline with RACS requirements, the following changes are suggested:

- Operation to modify existing VPrP must be added (e.g. `modifyVPrP()` method).
- Asynchronous messaging for notification of network changes and for feedback of admission control messages (e.g. `IpAppVPrN` and `IpAppVPrP` callback interfaces).
- Multicast element must be added to `TpPipeQoSInfo` datatype.
- A priority field must be added for each relevant element inside `TpPipeQoS Info` and `TpValidifyInfo` datatypes.

Further to these modifications to the OSA/Parlay Connectivity Manager APIs, we suggest that the QoS control capabilities be specified additionally as a Parlay X Web Service since it may facilitate adoption through better ease-of-use by application developers. Our new Parlay X Web Service for QoS control is detailed in the following section.

3   A New Parlay X Web Service for QoS Control

3.1   Service Description

The Parlay X Web Services model adheres to current best practices for Web Services technology. It provides discovery and access of services through a standard Universal Description, Discovery, and Integration (UDDI) registry, and considers security precautions in a manner prescribed by the WS-Security standard. Consequently, Parlay X does not require explicit Framework interfaces like in OSA/Parlay. Parlay X Web Services also provides abstracted and simplified functionality compared to OSA/Parlay SCFs, effectively making them easier to use for IT application developers.

In this section we describe a new Parlay X Web Service, QoS Control, for creating and managing a guaranteed traffic flow initiated by an application (third party QoS control). The overall scope of this Web Service is to provide functions to application developers to establish and maintain QoS pipes in a simple way. Using the QoS Control Web Service, application developers can invoke QoS handling functions without detailed telecommunication knowledge.

Figure 1 shows a scenario using the QoS Control Web Service to handle third party QoS control functions. The application is a portal for online videos (e.g. VoD), and invokes a Parlay X interface to initiate a guaranteed traffic connection between the closest video stream server or proxy (source) and video stream client (destination).

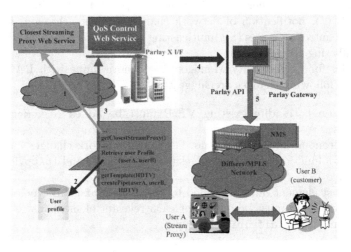

Fig. 1. Parlay X QoS Control scenario

In the scenario, when a customer chooses to watch a high definition video stream (e.g. on-demand movie, sports event, music video, or television show), the client application locates the closest video stream server or proxy (1) and (2). The application then retrieves the negotiated traffic conditions stored in a template (HDTV) and invokes a guaranteed traffic connection between the stream proxy and the customer who requested the video stream. After invocation (3) by the application, the QoS Control Web Service invokes a Parlay API (4) method using the OSA/Parlay SCS-CM (Service Capability Server – Connectivity Manager) interface. This SCF handles the invocation and sends a message (5) to a network management system (e.g. Alcatel 5750 Subscriber Services Controller) to configure the guaranteed path between user A and user B. The implementation of the network management system is beyond the scope of this document. It is assumed that the underlying service capabilities for controlling QoS is provided by the network management system, (a.k.a. resource controller, bandwidth broker, etc).

In an alternative scenario, the OSA/Parlay API interaction involving steps (4) and (5) could be replaced with a direct interaction between the QoS Control Web Service and the broadband network.

3.2 Specification

Parlay X Web Service specifications are decomposed into several documents and expressed using Web Services Description Language (WSDL) [13].The draft WSDL files (work in progress) for our QoS Control Web Service may be downloaded by contacting the author. Some comments on the *XML Schema data type* document and *Web Service interface* documents are described next.

XML Schema data type document. This document defines datatype definitions for QoS parameters, PipeInformation and various feedback messages. We have converted the existing Connectivity Manager datatypes into XML Schema *complex types* for the current version of the Web Service, but will consider using alternate datatypes when we become more familiar with using it to develop applications.

Web Service interface documents. Each interface is defined in a separate document containing the message and portType definitions. There are two interfaces defined, the first is the main QoS Control service interface and the second is for asynchronous application callback. We found that the need to navigate multi-level menus in the existing Connectivity Manager API was cumbersome to use during application development, so we have provided a single-level interface to reduce complexity.

3.3 Prototype Implementation

Our prototype implements some of the changes described in the previous section, and incorporates it into a Parlay X Web Service. We demonstrate the practicality of our QoS Control Web Service by implementing a mapping of high-level operations into actual low-level commands that are issued to simulated Alcatel 7750SR [14] devices. Alcatel 7750SR is the state-of-the-art in service router technology, featuring advanced QoS and VPN capabilities. Although the 7750SR has many other features, we are concentrating on the QoS control aspect in this paper.

7750 SR routers use QoS policies to control how QoS is handled at distinct points in the service delivery model within the device. There are different types of QoS policies that cater to the different QoS needs at each point in the service delivery model. QoS policies define classification rules for how traffic is mapped to queues; the forwarding class queues, queue parameters used for policing, shaping, and buffer allocation; and QoS marking/interpretation.

If the service core network is oversubscribed, a mechanism to traffic-engineer a path through the core network and reserve bandwidth must be used to apply strict control over the delay and bandwidth requirements of high-priority traffic. In the 7750 SR, Resource Reservation Protocol with Traffic Engineering (RSVP-TE) can be used to create a path defined by a Multi Protocol Label Switching

(MPLS) Label Switched Path (LSP) through the core. Premium services are then mapped to the LSP with care exercised not to oversubscribe the reserved bandwidth. If the core network has sufficient bandwidth, it is possible to effectively support the delay and jitter characteristics of high-priority traffic without utilizing traffic engineered paths, as long as the core treats high-priority traffic with the proper Per Hop Behavior (PHB).

We used simulated Alcatel 7750SR devices, which meant that the actual configuration of network QoS policies is not functioning yet, but are simulated with text output messages indicating the parameters that would be sent to the network element. We are not testing the performance of the 7750SR's switching capabilities. The performance of the 7750SR has already been documented in independent third-party reports including [15].

It must be noted that the 7750SR in conjunction with 5620 Service Aware Manager or Alcatel 5750 Subscriber Services Controller already opens the management modules for an OSS application via an XML interface. This OSS interface allows provisioning of services and policies, integration into existing multi-vendor OSS systems, fault management, equipment and inventory management. However, this 'open' interface is only useful for the operator who wishes to integrate the management functions of the 7750 SR into an OSS. It does not open up the network functions to third party service providers who could use these new functions in creating innovative new applications.

The value of our contribution is that by utilizing standardized Parlay X Web Services, it is possible to open up the advanced 7750SR functions such as VPNs and QoS policies to an increasing number of applications, while keeping third party access manageable through UDDI and WS-Security mechanisms.

3.4   Example QoS-Aware Multimedia Application

The purpose of providing an example QoS-aware application is to demonstrate that service providers could easily and practically use this Parlay X Web Service as a part of their applications in order to make the most out of the functionality of next generation networks. These service capabilities that were previously inaccessible have been packaged and presented in a fashion that is open, practical, and manageable.

The application scenario is an advanced entertainment portal / enhanced interactive broadband digital television with instant video on demand. It is a fully converged video system with additional features such as electronic program guide reminders via phone and text messaging. The QoS Control Web Service is essentially used to provide feedback to the application so that the application may respond as it wishes. The application not only uses the proposed Web Service, but also takes advantage of the existing Third Party Call control and Short Messaging Web Services.

After the service is authenticated and registered with the UDDI registry, the application is able to discover and request access to the QoS Control Web Service. In our example QoS-aware multimedia application, QoS and bandwidth parameters can be requested and modified on demand by the application in a very practical way. Additionally, the application is able to react to feedback and notification messages provided by the Web Service. The application is shielded from the complexity of the underlying protocols and is able to utilize the benefits of QoS through an easy to use Web Service.

At the same time, the application could also access other Web Services such as Third Party Call control and Short Messaging. By including this new QoS Control Web Service, the service provider is able to provide a value-added service that takes advantage of convergence between different networks. The user could be provided with a service that operates transparently between IP, PSTN and mobile technologies. For example, if the application is unable to establish a QoS pipe at the time due to insufficient resources. It could send the user an SMS when resources become free, or make a call to the user through VoIP / mobile or fixed line.

4   Summary and Future Work

We gave an overview of the OSA/Parlay Connectivity Manager for controlling QoS in telecommunications networks, and described how this standardized API falls short of satisfying the requirements from ETSI TISPAN RACS drafts. The following suggestions were made to improve the existing OSA/Parlay Connectivity Manager: operation to modify existing VPrPs must be added; asynchronous messaging for notification of network changes must be added; asynchronous messaging for feedback of admission control results must be added; multicast element must be added to the QoS parameter description datatypes; and priority fields must be added for each relevent element inside the QoS parameter description datatypes. Next, we described our new Parlay X Web Service for QoS control, which incorporates the suggestions to Connectivity Manager, as well as provides a simplified abstraction for better ease-of-use by application developers. We have implemented a basic prototype of this new QoS Control Web Service, and described an example QoS-aware multimedia application.

We are still working on improving the prototype implementation of our new Parlay X Web Service and we are gaining more experience in using it to develop example applications. This learning-outcome process is helping us to evaluate and measure the effectiveness of our open system. In the measurement of open systems [16, 17], the major qualities which characterize an effective open system are the ease with which components can interoperate together (interoperability) and the ease with which components can be transported from one system to another (portability). Direct measurement of these characteristics is of course important to confirm that the system fulfils its open objectives, i.e. it supports interoperability and portability. However, such measures are of little use to the engineer designing an open system. One needs to build predictive measures; measures that will ensure interoperability and portability in the built system. To date, the best predictive measures available focus on how well the components of the infrastructure comply with well-established standards. We will continue our evaluation as a part of future work.

Acknowledgement

We acknowledge the funding by UTS, Alcatel and the ARC via grant LP 0219784, which includes an APA(I) for Samson Lee.

References

1. Moerdijk, A.J., Klostermann, L.: Opening the Networks with Parlay/OSA: Standards and Aspects Behind the APIs. IEEE Network (2003) 58–64
2. The Parlay Group: Parlay APIs 4.0 and Parlay X Web Services. Whitepaper (2002)
3. 3GPP: TSG Core Network and Terminals WG5.
   Technical report, http://www.3gpp.org/TB/CT/CT5/CT5.htm (2005)
4. ETSI TISPAN: Webpage. Technical report, http://portal.etsi.org/tispan/ (2004)
5. Lozinski, Z.: Parlay Product Catalog. Technical report (2004)
6. Turner, K., Magill, E.H., Marples, D.J.: Communications Services: The Technology of Call Control. Wiley, John & Sons, Incorporated (January 2002)
7. Jain, R., Anjum, F., Bakker, J.L.: Programming Converged Networks: Call Control in JTAPI, JAIN, and Parlay/OSA. Wiley, John & Sons, Incorporated (November 2003)
8. Golding, P.: Next Generation Wireless Applications. Wiley, John & Sons, Incorporated (May 2004)
9. Joint Working Group: Barcelona Meeting Report. Technical report,
   http://www.3gpp.org/ftp/tsg_cn/WG5_osa/TSGN5_29_Barcelona/Docs/N5-040708%20DRAFT%20A%20Report_CN5_29_Barcelona.zip (2004)
10. Lee, S., , Leaney, J., O'Neill, T., Hunter, M.: Open API for QoS control in Next Generation Networks. In: Asia Pacific Network Operations and Management Symposium (APNOMS'05). (2005)
11. The Parlay Group: ETSI ES 202 915-10 OSA API Part 10: Connectivity Manager SCF. Technical report (2004)
12. Sheridan-Smith, N., Leaney, J., O'Neill, T., Hunter, M.: A Policy-Driven Autonomous System for Evolutive and Adaptive Management of Complex Services and Networks. In: 12th Annual IEEE International Conference and Workshop on the Engineering of Computer Based Systems (ECBS), Greenbelt, Maryland, USA (2005)
13. W3C: Web Services Description Language (WSDL) 1.1. Technical report, W3C (2001)
14. Alcatel: Alcatel 7750SR Service Router Product Literature. (2005)
15. Shippam, P., Ridgewell, P.: Next-Generation Routers: A Comprehensive Product Analysis. Technical report, Heavy Reading (2004)
16. O'Neill, T., Rowe, D., Leaney, J.: An Open Computer Based System (CBS) Quality Metrics Framework. In: 1998 IEEE Conference and Workshop on Engineering of Computer-Based Systems (ECBS'98), Jerusalem, Israel (1998) pp158–165
17. Leaney, J., Rowe, D., O'Neill, T.: Issues in the construction of new measures within the discipline of Open System. In: 9th Asia-Pacific Software Engineering Conference, Gold Coast, Australia (2002)

Remote Service Invocation Through Heterogeneous Networks Using Open Environments

Alejandro Bascuñana Muñoz[1] and Tomás Robles Valladares[2]

[1] Ericsson España S.A, Vía del los poblados 13, 28033 Madrid, Spain
alejandro.bascunana@ericsson.com
http://www.ericsson.com
[2] Universidad Politécnica de Madrid, Paraninfo de la ciudad universitaria, 28040 Madrid, Spain
trobles@dit.upm.es
http://dit.upm.es

Abstract. Current OSA/PARLAY standard defines an architecture that enables service application located into an ASP to invoke network services capabilities through an open standardized interface. Nevertheless, architectures proposed within this standard work are isolated from other OSA/PARLAY infrastructures belonging to other network operators. In this research paper we deal with the invocation of network services located in visited networks from applications attached to a Home Network using OSA/PARLAY interfaces. We propose an extension to the OSA/PARLAY Framework signaling system so that it will work in a multi-OSA environment. This enhancement will require the definition of new interface classes among Frameworks in order to manage global service execution. The proposed method adds a new Framework-to-Framework interface that allows applications attached to a home network the access to service capability servers located in visited networks. In this enhanced OSA/PARLAY architecture, the applications may access remote services without integrating or dealing with any network negotiation.

1 Introduction

OSA specifications [1] define an architecture that enables service applications to access network functionalities through open standardized interfaces. Within this environment, the Applications access functionalities using the Service Capability Servers (SCSs). The OSA Framework is a general component supporting Services (Service Capabilities) and Applications providing Authorization, Authentication and Service Level Agreements (SLA) verification. According to this architecture, a registered Application may access SCSs according to SLAs previously negotiated and signed with the operator. Nevertheless, any Application that requires the use of SCSs provided by different operators has to register into a different Network Operators' Framework, sign the suitable SLAs, and assume all the complexity of contracts involving security and communications aspects, as well as the problem of selecting the suitable SCSs among all of those available. More unfortunate is that the Application has to foresee the usage of such remote SCSs due to unavailability of any automatic and real time mechanism for negotiating this usage if a previously signed contract between the Application provider and the Network Operator has not been introduced. This interconnection will be the base of the so-called B3G/4G networks.

This paper proposes extensions to the OSA/PARLAY architecture in order to allow Applications to use Services provided by SCSs located in a visited network. This ex-

T. Magedanz, E.R.M. Madeira, and P. Dini (Eds.): IPOM 2005, LNCS 3751, pp. 39–48, 2005.

tension requires the definition of a new API for the interconnection of OSA/PARLAY Frameworks with the deployment of the corresponding protocols. This new F2F interaction requires the definition of the suitable SLA between Domains, which have to unite with the currently standardized SLAs. Under this new context, access constraint evaluation requires the definition of a new mechanism for combining dissimilar SLAs involved in the use of a SCS by an Application in a typical case.

Section 2 outlines the essential characteristics of OSA/PARLAY Architecture; Section 3 describes the proposed solution for the problem of remote service invocation; Section 4 provides details about the Service Level Agreement (SLA) modifications required by the solution proposed in section 3; Section 5 describes the central characteristics of the prototype; and Section 6 summarizes the fundamental conclusions of this paper.

Two use cases can be used as an example to illustrate how a lack of interconnection of OSA/PARLAY Domains limits the use of basic services on the OSA/PARLAY architectures: An international parcel company tracking their trucks, not only in its country but also in other countries, and an international healthcare company X tracking their heart patients, not only through the wide area that its home network operator covers, but also within indoor spaces, e.g. within huge airports using Bluetooth positioning technology [7] or within underground facilities without a Public Land Mobile Network (PLMN) Operator but with WLAN coverage.

2 OSA/PARLAY Architecture

OSA specifications define an architecture that enables service application developers to make use of network functionalities through an open standardized interface. The OSA API as defined in [1] is split into three types of interface classes [4] (figure 1):

- Interface classes between Applications and Framework that provide.
- Interface classes between Applications and Service Capability Features (SCF).
- Interface classes between the Framework and the Service Capability Features.

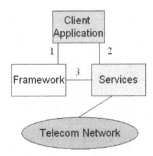

Fig. 1. Standard interfaces within the OSA/Parlay architecture

Today, applications using network services interface telecom networks by means of Intelligent Networks (IN[4]) or Customized Application for Mobile network Enhanced Logic (CAMEL[5]) services, but these networks are not open to external developments outside the operator, leading to a "wall-garden" scenario, even in the cases of third-party application providers

The current OSA/PARLAY architecture (version 4.1 [1]) is defined as a stand-alone domain. Frameworks specified by the OSA/PARLAY recommendations are not allowed to have communication with other OSA/PARLAY systems deployed in other Telecom Networks. An operator will have the service provider role and hundreds of services will be deployed each year. Millions of users will use these services; the technology on the operator side should permit the signing of agreements among them, just for allowing ASPs to access services located in other networks which differ from the contracted network. The only way of handling the interconnection among different domains is to have new contracts with all of the Network operators of countries that have foreseen user locations. This is clearly in contrast to GSM whose roaming services are negotiated between operator and the end user, who uses the roaming service only supported by the contract with a Home Network Operator.

There are also European Information Society Technologies (IST) projects working on the service-roaming problem using an OSA/PARLAY approach. Open Platform for Integration of UMTS Middleware (OPIUM [6]) proposes the use of middleware for performing support of a multiple service registration approach. This approach only simplifies the operation and maintenance of multiple services, but fails to create a real framework for service sharing, interconnections agreements, and transparent services access by the end-user.

3   Remote Service Invocation in an OSA/PARLAY Environment

In order to allow end-users to access Applications located in remote networks, we propose the interconnection of OSA/PARLAY domains. Attending to the current architecture, the key element we identified for supporting any interconnection is the OSA/PARLAY Framework. Our proposal is based on the definition of the so-called Virtual Global Framework (VGF). VGF is defined by the interconnection of different OSA/PARLAY Frameworks, using suitable protocols and rules by the corresponding SLAs, in order to allow any application to access local or remote services transparently. Figure 2 shows a general model for a global service provision network where different Frameworks are interconnected, and the applications register to any of them with the SLA. In this model the VGF allows applications to access services distributed along several core networks. In order to complete the definition of VGF, we have to enrich current OSA/PARLAY definitions with the definition of a Framework-to-Framework (F2F) interface, the corresponding protocol for performing the interconnection and mechanisms for defining, managing, and negotiating SLAs between two Domains.

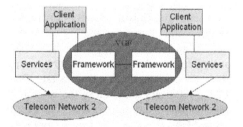

Fig. 2. F2F Interface and VGF

With the introduction of VGF, operators could offer their services not only to their local customers, or ASPs, but also to ASPs located in other networks with previously signed SLAs and contracts. The VGF concept allows operators to offer services beyond their limits and allows their ASPs to offer their applications to multinational customers (See examples provided in the introduction). Client Applications gain access to remote services based on contracts with local Network Operator and the interconnection agreements among different Network Operators.

The definition of a F2F interface as can be seen in Figure 2, together with the correct definition of the SLAs, will allow remote service execution and service network roaming in an OSA/PARLAY environment.

The VGF may be implemented following two different strategies during the registration phase between the frameworks:

- Off-line mode
- On-line mode

The key difference between On-Line and Off-Line protocols is the management of the communications between frameworks. While On-Line communication executes a verification of service level agreements and service availability in the visited network in real time, Off-Line communication updates SLAs and available services in a non-real time method, updating information as it changes. Let's analyze each alternative in detail.

Off-Line Scenario

Off-Line scenario should be applied to these VGF positioned in a stable environment that suffers only sporadic changes. Prior to an Off-Line access, the interconnecting frameworks must perform several actions dealing with: Registration, Service Availability, Advertisement, and SLA Negotiation.

Those processes between frameworks can be grouped into 3 basic steps:

1. Advertise the existence of a new framework (Remote) that can be accessed by the operator framework that has the application (Local).
2. Publish interfaces that will facilitate local client applications to consume Local Framework resources, which in fact are deployed as Remote Framework services.
3. Authorize service level agreement negotiation between the two frameworks.

The new Framework references and the remote available services are stored by the framework. A fourth step is completed when the framework adds or changes framework services, or when modifications are introduced in the SLA. The corresponding updated information will be interchanged between both associated frameworks.

Figure 3 shows the overall process, from the authentication phase to the services access phase in a remote framework (Visited network operator).

In a first step of the F2F communication (see Figure 3), frameworks authenticate each other (steps 1 and 2), and once authorization has been granted, then service availability and previously negotiated services level agreements, both F2F and offered services, are registered on the correlating frameworks (step 3). Once these tasks have been completed, any of the Fw peers could close the session. At the end of this first phase, the Fw stores information regarding remote services available from this remote Fw and the corresponding SLAs for use when local applications request such services (step 4).

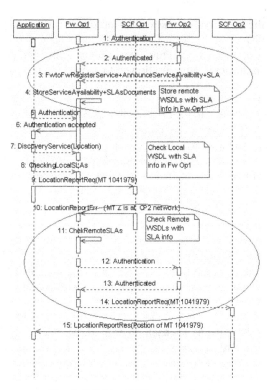

Fig. 3. Off-Line F2F communication

When an application wants to use a service (i.e. Location Services), the first stage follows the OSA/PARLAY standardized flow: Authentication in the Framework (steps 5 and 6), Discovery of the service (step 7), corroboration of Service Level Agreements (step 8) and service accessing (step 9). If at the end of this operation, the service returns an error to the framework (this step requires changes in the OSA/Parlay specifications to support this scenario) due to the mobile device being located in a foreign network (step 10), then start-up of the off-line remote service is executed, using the information stored at the end of the first step of the Fw check that the stored SLA applies to such remote service (step 11). The Fw then attempts to access the service located on the visited Network Operator on behalf of the application (steps 12, 13, 14 and 15).

In interaction 10, the addressee of the message LocationReportErr has been changed and redirected to the FW1 to allow applications to be integrated within an offline federation environment without adding new functionalities to them. Instead, the FW Op1 needs to have implemented the IpAppUserLocation interface, and SCF Op1 needs to redirect its locationReportErr messages there. This new functionality must be added to the framework and SCSs in order to implement this scenario.

In figure 3 proposed changes to be introduced in the OSA Parlay specifications to support this scenario have been remarked within a red ellipse.

The framework corroborates using its SLAs database, if the service properties requested by the application are authorized by both the visited location service and the

F2F SLA. All of these tasks are done within the framework without establishing any contact with the visited framework. Once SLAs have been verified and the service discovery has been performed, the Frameworks reciprocally authenticate, interchange security tokens, and the application is allowed to access the remote service execution.

On-Line Scenario
In an On-Line environment the service and SLA verification are performed in real time whenever a remote service is requested. In this scenario there is no previous negotiation between frameworks. The application wants to use a service and doesn't know in which network the Mobile Wireless Device (MWL) is located. In the sequence diagram of Figure 4, a localization service is accessed.

In the On-Line case, a previous Authentication between Frameworks does not apply. The communication and SLAs interchange is done in real time by the following process:

1. An Application tries to access in order to locate a mobile device using services provided by the local framework. Follows the Standard. (step 1,2,3,4,5)
2. Once the location service informs of an error to the Fw (Step 6) as a result of the mobile device being located in a foreign network, the on-line procedure is then started.
3. The Home Fw attempts to use the location service located in the visited network, so the Frameworks authenticate each other in the Visited Network Operator (steps 7, 8). If this procedure is successful, then F2F and service specific SLAs are verified in an on-line modal existence against information stored in the visited domain database.

If the application solicits a service within the limits of the SLA, then the remote service invocation will be accepted (step 10), otherwise, this remote service invocation will be rejected.

The pivotal advantage of this procedure is that the application only contacts with its local framework when it wants to access a service, and the framework manages the following process along with the relationship with other federated OSA/PARLAY environments. The application is only registered in one framework and does not need to be registered in all the associated domains.

4 SLA in the F2F Communication

SLAs are key elements in the management of relations between players (Network Operators, ASPs and end-users) involved in a distributed and collaborative service environment. Furthermore, SLAs define access rights for each of the specified elements. For each of the elements that compound SLAs, the following types of present values could be defined: Discrete value ranges, continuous value ranges and enumeration list.

OSA/PARLAY standards define the procedures required to manage SLAs and to regulate the relationship between Network Operators (Services) and Application Providers (Application). SLAs between Services and Applications are signed by Network Service Operators (Framework) and Applications Providers. This document is stored in the service properties area within the Framework and has the following SLA document signing process: Once the Application discovers the service it wants to use, the

Framework sends the Application a service agreement text that will be signed by the Application and later by the Framework.

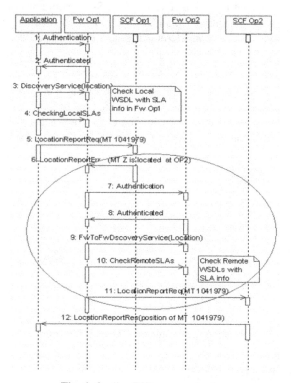

Fig. 4. On-line F2F communication

In addition to the F2F interface definition, it is extended to the level of relationship management among the players by adding two additional SLAs contracts (Figure 5) such as:

- A general F2F (or service domain to service domain relationship) SLA. This SLA will be used for managing relationships among network operators involved in a service remote access.
- A more focused SLA to manage access to services located in a remote service domain. This SLA manages the relationship between Applications and remote services and defines the quality values that a foreign application can use to access the service.

Several SLAs are needed to define how an application accesses a remote service. In the proposed model, the Framework automatically combines these SLAs. When an application wants access to a remote service, it is necessary to manage a combination of general F2F SLAs with a more specific service oriented SLA, both in the on-line and off-line modes. Occasionally a decision from several services offering different SLAs could be requested by the application.

In order to structure an automatic mechanism to manage the SLA combinations proposed above, a group of operations algorithms for processing support of a given

SLA document has been prepared. Operations are defined to work with sets of SLAs. These operations are Addition, Maximum, Minimum, Average, Union and Intersection. Basic operations define how different sets of SLAs could be combined.

In order to manage this situation, an SLA combination function has been defined. The F function combines several SLAs using operations to produce a single SLA as result.

$$SLA = F (SLA1, SLA2, ..., SLAn).$$

The output of this function generates for each of the parameters included within SLA1, SLA2, SLAn etc., a new value based on the type of value and the type of SLA signed between Operators. In out proposal the Ó function is hard-coded into the Fw, but the definition of dynamically reconfigurable SLA brokers is the next step.

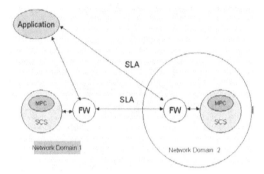

Fig. 5. SLAs within the Global architecture

5 Prototype Implementation

In order to illustrate the feasibility of the proposed model, we developed a prototype that illustrates both interconnection modes and allows a location application to use position services of remote OSA/PARLAY domains. The prototype over the Java API for Integrated Networks (JAIN SPA [8]) implementing:

- Basic functionalities of the OSA/PARLAY architecture according to current standards.
- Modification in the Framework code for support of new functionalities.
- A SCS providing a mobility service. This service has been provided by the use of the MPS. The MPS is a mobility simulator developed by Ericsson [9].
- An Application use the mobility service for locating mobiles.

Figure 6 depicts the simulated scenario, where a location application connected to DOMAIN 1 is able to locate a mobile equipment (in our case places on a car) that moves from the area covered by the operator of Domain 1, to the area covered by the operator of Domain 2.

The implementation of the prototype was performed in two phases. During the first phase, we developed a basic OSA/PARLAY architecture which included a Framework and a SCF based on the Ericsson MPS. In this phase, we also developed the Location Application and managed a SLA between the Location Application and the Local Framework.

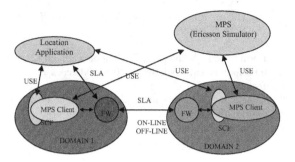

Fig. 6. Prototype Scenario

During the second phase, we implemented on-line and off-line modes and provided support for SLA negotiation between the two frameworks. The application has not been modified, which confirms the transparency of the proposed enhancement. It is important to note that although we worked over a basic version of the Framework functionalities, we only have to increment the size of the code by 45%, and that no other element or functionality of the original OSA/PARLAY architecture has been modified, but merely extended, thereby preserving backward compatibility in all the elements. Looking into the code, we can summarize that half of this increment is due to the management of the SLA and the other half is due to the interconnection and remote access.

6   Conclusions

With the introduction of the VGF, operators could offer their services not only to their customers or ASPs, but also to ASPs located in other OSA/PARLAY networks with previously signed SLAs and contracts. The VGF concept allows operators to offer their services beyond their limits and allows their ASPs to offer their applications to international customers.

A key element of this proposal is the management of SLAs, which involves definitions of new SLAs, procedures for management and negotiation of SLAs, and the definition of a general mechanism for combining different SLAs in order to evaluate which services the applications may use.

This proposal has been validated by the implementation of the prototype. The prototype demonstrates that only minor and local modifications were required for adapting JAIN reference implementation to our proposal. Modifications are focused on current OSA/PARLAY Framework element, and most of them refer to the management of SLAs.

References

1. Parlay 4.1 Specification. http://www.parlay.org/specs/index.asp
2. IEEE Network Magazine. May 2003. Vol 17.N°3. Opening the networks with Parlay/OSA: Standards and aspects behind the APIs.
3. TSI ES 201 915-1 v1.4.1 (2003-07). Open Service Access(OSA); Application Programming Interface (API); Part 1: Overview (Parlay 3)

4. Intelligent Networks. The path to global networking. *P.W. Bayliss* (AT&T Network Systems, USA) 1992. ISBN: 90 5199 091 X
5. 3GPP, TSG CN WG2
6. OPIUM. Open Platform for Integration of UMTS Middleware. http://www.ist-opium.org/
7. Bluesoft INC. 1450 Fashion Island Blvd. Suite 510 San Mateo, CA 94404. USA. http://www.bluesoft-inc.com
8. JAIN. Java API for Integrated Networks. http://java.sun.com/products/jain/
9. Ericsson Mobility World. http://www.ericsson.com/mobilityworld/

Modeling of Dynamic Pricing by Market Demand in Multiple QoS Networks

Sang Ki Kim[1,2] and Mun Kee Choi[2]

[1] ETRI, BcN Research Division, Kajung-dong 161 Yusong, 305-600 Daejeon, Korea
`kimsang@etri.re.kr`
[2] ICU, School of IT Business, Munjee-dong 103-6 Yusong, 305-732 Daejeon, Korea
`mkchoi@icu.ac.kr`

Abstract. In order to manage congestion problems and allocate network resources, many researchers have studied Internet pricing over the last decade. However, much of their research results have been limited by their reliance on the over-simplified demand model, and are not intended for adaptation to emerging multiple class environments such as the Diffserv network. For example, user utility is generally represented by a logarithmic form that is related to unit elasticity demand, but is not effective in representing user demand in the real Internet service market. We extend a dynamic pricing scheme by generalizing a demand model and applying it to the multi-class Diffserv network; and develop a simulation framework to compare the engineering and economic performance of our dynamic pricing model to those of static pricing.

1 Introduction

Flat-rate pricing, which is based on the best-effort service model, has contributed to the explosive growth of Internet usage. However, this current pricing method has confronted with the moral hazard problem referred to as the "tragedy of commons" since there is no incentive for users to reduce network usage when congestion occurs.

In order to manage congestion problems and allocate network resources fairly, many researchers have studied Internet pricing over the last decade. Most research on Internet pricing has been focused on dynamic pricing, which entails increasing usage price when the network becomes congested, and decreasing the price when congestion is lessened.

Representative dynamic pricing proposals include MacKie-Mason and Varian's Smart Market [2], Priority Pricing from Gupta et al. [3], Proportional Fair Pricing (PFP) from Kelly et al. [7, 8], Dynamic Capacity Contracting (DCC) from Kalyanaraman et al.[6], and Wang and Schulzrinne's Resource Negotiation and Pricing (RNAP)[14, 15].

However, most of these methods are limited by their reliance on an over-simplified demand model, and are not intended for adaptation to emerging multiple class environments such as the Diffserv network. For example, user utility is generally represented by a logarithmic form or its variants. Price elasticity of demand in this utility function is derived where users have unit elastic demands. In this case, however, the user is assumed to pay always the same expenditure to the network at any price level

T. Magedanz, E.R.M. Madeira, and P. Dini (Eds.): IPOM 2005, LNCS 3751, pp. 49–57, 2005.

of the bandwidth. This assumption is too restrictive when we consider real situations. In the monopoly market there are two important aspects of demand behavior: linear demand and constant elasticity demand. Unit elasticity demand is a special case of the constant elasticity demand model, but it is not well-suited for representing user demand in the Internet bandwidth market.

In [15], Wang and Schulzrinne suggested a complete framework for dynamic pricing on Diffserv networks. Even though their approach is likely to fit better than other researches on the evolving Internet in a Diffserv environment, their model is based on a simple logarithmic utility function related to unit elastic demand.

In this paper we extend the dynamic pricing scheme by generalizing the demand model and applying it to multi-class Diffserv networks. We also develop a simulation framework to compare the engineering and economic performance of our dynamic pricing model to those of static pricing.

The paper is organized as follows. In Section 2 we develop our pricing framework to represent provider decision making on pricing in single class environment. We extend the result of Section 2 into multiple classes situation in Section 3. Section 4 describes our experimental simulation and its results. Finally, in Section 5 we present our conclusions.

2 Dynamic Pricing for Single Class Environment

Following the work of the pricing and flow optimization model for congestion control [8, 9], we investigate the pricing model in terms of maximizing the aggregated utility function.

Let's consider a network where n users are using a network which has a resource capacity of c. In our basic model, the capacity of network resources represents bandwidth of link. For each user i, x_i is the amount of bandwidth given to the user and $u_i(x_i)$ is the utility function of user i.

Then the optimization model of social welfare, which is aggregated users' utilities in our model, can be represented as

$$\max \sum_{i=1}^{n} u_i(x_i)$$
$$s.t. \sum_{i=1}^{n} x_i \le c$$

(1)

To maximize aggregated users' utilities, all network resources should be distributed to users since there are strictly increasing utilities.

In most Internet pricing literatures, the utility function is in a logarithmic form thanks to its simplicity.

$$u_i(x_i) = w_i \log x_i .$$

(2)

where w_i is willingness-to-pay of user i.

By applying the Lagrangian method to the logarithmic utility case, we can get the following solutions:

$$p = \frac{\sum w_i}{c}$$

$$x_i(p) = \frac{w_i}{p} = w_i p^{-1}$$

(3)

Note that p is bandwidth price.

The result shows us that the maximization of aggregated user utilities can be done by setting the resource price according to total willingness-to-pay and allocating the resource to the user in proportion to their willingness-to-pay. From the viewpoint of market mechanism, the price is set by demand and supply. The demand is the sum of whole willingness-to-pay and the supply is the amount of resource.

User demand of the bandwidth in the second equation of (3) can be generalized to the constant elasticity demand function in microeconomics, which has the same elasticity at all prices as shown in the following form [10, 11]:

$$x(p) = a\, p^\varepsilon .$$

(4)

Price elasticity of demand in (3) is -1, which means users have unit elastic demands. So, (4) is a more generalized form for price elasticity of demand.

When comparing (3) and (4), we are assured that the demand function of (3) is a special case of the constant elasticity demand function with the following relations:

$$a = w$$
$$\varepsilon = -1 .$$

(5)

If users do not have logarithmic utility but multiplicative utility as follows, the price elasticity of demand would not be -1.

$$u_i(x) = w_i x^\alpha .$$

(6)

Note that the socially optimal price and utility function have the following relationship [11]:

$$p = u'(x) .$$

(7)

So, we can yield the price and demand function from a derivative of utility function (7).

$$u'(x) = w\alpha x^{\alpha-1} = p$$

$$x = (w\alpha)^{\frac{1}{1-\alpha}} p^{\frac{1}{\alpha-1}} .$$

(8)

Then, we can derive the following relations from (4) and (8):

$$\varepsilon = \frac{1}{\alpha-1}$$

$$a = (w\alpha)^{\frac{1}{1-\alpha}} = (w\alpha)^{-\varepsilon} .$$

(9)

$$p = w\alpha x^{\frac{1}{\varepsilon}}$$

From the above equations, we can derive a new demand function with price elasticity as follows:

$$x(p) = (w\alpha)^{-\varepsilon} p^{\varepsilon}. \tag{10}$$

So, the demand function (3) from logarithmic utility is regarded as a special case of the demand function (10) with multiplicative utility function when $\varepsilon = -1, \quad \alpha = 1$.

Now let's consider the optimization model to maximize provider revenue, which is given as:

$$\begin{aligned} Max \quad & \sum xp \\ s.t. \quad & \sum x \le c \end{aligned} \tag{11}$$

By applying the Lagrangian Multiplier method, we can get the following solution:

$$\begin{aligned} c &= \sum (w\alpha)^{-\varepsilon} p^{\varepsilon} \\ p &= \sum (w\alpha) c^{1/\varepsilon} \end{aligned} \tag{12}$$

As the price elasticity of demand should have a non-positive value, we can rewrite it with an absolute value to clarify the meaning of the result.

$$\varepsilon = -|\varepsilon|. \tag{13}$$

If we adopt the above notation, the optimal price and user demand is as follows:

$$\begin{aligned} p &= (1 - \frac{1}{|\varepsilon|}) \frac{\sum w}{c^{1/|\varepsilon|}} \quad if \ \varepsilon \ne -1 \\ &= \frac{\sum w}{c} \quad if \ \varepsilon = -1 \\ x &= \left[\frac{w}{p}(1 - \frac{1}{|\varepsilon|}) \right]^{|\varepsilon|} \quad if \ \varepsilon \ne -1 \\ &= \frac{w}{p} \quad if \ \varepsilon = -1 \end{aligned} \tag{14}$$

Here we can reconfirm that the basic model is a special case of the enhanced model, where $\varepsilon = -1, \quad \alpha = 1$.

3 Dynamic Pricing for Multiple Classes Environment

In a Diffserv network, each router manages the bandwidth of its links and buffer space for separated service classes. The unit price of link bandwidth in our model depends on the kind of service class and the level of congestion of the network.

Let's assume there are K classes in a network, and the unit price of bandwidth for class k is p_k. Then the core of the network provider's problem is determining the optimal unit price of bandwidth to maximize revenue. Revenue maximization on the Differserv network can be expressed as follows:

$$\max \sum_{k=1}^{K} x_k(p_k)p_k$$
$$s.t. \sum_{k=1}^{K} x_k(p_k) \le C$$
(15)

In the Diffserv model, a class with lower load leads to lower expected delays. A higher level of service class is expected to have a lower average load, and hence lower average delay. So we need to charge services proportional to its expected load to reflect the cost of different quality of service.

Assume that a basic unit price of bandwidth is p_{basis} if all its bandwidth is used. If the expected load ratio of service class k is ρ_k, the basic unit price of bandwidth for class k is as follows[12, 15]:

$$p_k = \frac{p_{basis}}{\rho_k}.$$
(16)

Let's assume the user utility function of each class uses the same kind of utility model. Then we can consider the user problem of each class independently with the other classes. The optimal demand function for bandwidth of class k was derived as follows from equation (14) of the previous section. Note that x_k and w_k in the following equation are aggregated demand and willingness-to-pay for class k, while those of the previous section are for each user.

$$x_k(p_k) = \frac{w_k}{p_k} \quad if \ \varepsilon = -1$$
$$= \left[\frac{w_k}{p_k}\left(1 - \frac{1}{|\varepsilon|}\right) \right]^{|\varepsilon|} \quad if \ \varepsilon \ne -1$$
(17)

However, the real bandwidth consumption of class k should be $\tilde{x}_k = \frac{x_k}{\rho_k}$, as the network should keep the load ratio for the quality of service class.

Then we can rewrite the network provider's optimization model as follows:

$$\max \sum_{k=1}^{K} x_k p_k$$
$$s.t. \sum_{k=1}^{K} \frac{x_k}{\rho_k} \le C$$
(18)

The constraint of the above model comes from $\sum \tilde{x}_k \le C$.

The Lagrangian of the maximization model with multiplier μ and its solution can be derived as:

$$Z = \sum x_k p_k + \mu(C - \sum \frac{x_k}{\rho_k})$$

$$\mu = p_{basis} \tag{19}$$

$$C = \sum \frac{x_k}{\rho_k}$$

If the utility function is non-logarithmic, the optimal price for basic unit and that of class are as follows:

$$C = \sum \frac{1}{\rho_k} \left[\frac{\rho_k w_k}{p_{basis}} \left(1 - \frac{1}{|\varepsilon|} \right) \right]^{|\varepsilon|}$$

$$P_{basis}^{*\,|\varepsilon|} = \frac{\left(1 - \frac{1}{|\varepsilon|} \right)^{|\varepsilon|}}{C} \sum \frac{\rho_k^{|\varepsilon|} w_k^{|\varepsilon|}}{\rho_k} \tag{20}$$

$$P_{basis}^{*} = \left(1 - \frac{1}{|\varepsilon|} \right)^{|\varepsilon|} \frac{\sum \rho_k^{\left(1 - \frac{1}{|\varepsilon|}\right)} w_k}{C^{\frac{1}{|\varepsilon|}}}$$

$$p_k^{*} = \frac{\sum w_k}{C \rho_k}$$

4   Experiments by Simulation

In order to validate our dynamic pricing model, we used ns-2 (network simulator version 2) to simulate a Diffserv network environment. Figure 1 shows our network topology for simulation experiments which consists of two core routers and six edge routers. Several tens of hosts are connected to each edge router. The links connecting the core routers are 100Mbps, the links between core and edge routers are 50Mbps, and the links between edge routers and end hosts are 20Mbps. Each router has a WRR (Weighted Round Robin) scheduler. Congestion is simulated between the two core routers.

In the experiments, we compare the static pricing (SP) policy and the dynamic pricing (DP) policy from the viewpoint of performance and efficiency. SP policy applies a constant price rate for each class without considering the variation of demand; while DP policy estimates the demand of users' willingness-to-pay for each period in order to set the dynamic price rate for each class.

The main focus of the experiments is the comparison of engineering performance and economic performance between SP and DP policies when a congestion situation occurs. We analyze performance characteristics such as packet loss and link utilization, and compare them for engineering efficiency. For economic efficiency, user surplus and provider revenue are selected as comparison countermeasures.

Figure 2 shows a comparison of static and dynamic pricing from the viewpoint of engineering performance, which consists of packet loss and resource utilization for given willingness-to-pay (WTP). For static pricing, 3 levels of base prices were experimented: 10, 15, and 20. The expansion of average WTP led to an increase in entering packet traffic and a rise in resource utilization, which then increased the occupation ratio of link bandwidth in turn. When the transmission rate goes beyond the threshold of link bandwidth, packet loss occurs. In the static pricing policy, there is no disincentive for network users to discourage them against increasing the traffic rate. So the more the user has the willingness-to-pay, the higher the network suffers from overload situations. However, utilization and packet loss are stable in dynamic pricing. The adjusted price in the dynamic scheme prevents people from sending excessive traffic to the network.

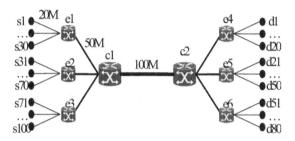

Fig. 1. Experimental Network Topology

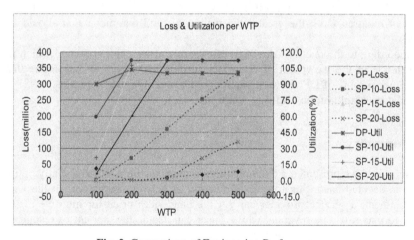

Fig. 2. Comparison of Engineering Performance

Figure 3 shows consumer surplus and provider revenue according to WTP. In this result, the consumer surplus is calculated by subtracting payment from user utility in which we consider only transmission bandwidth. This means that QoS-related disutility such as packet loss and queuing delay in the buffer space is not considered for surplus value. The comparison result between SP and DP is dependent on the level of static price and WTP. While consumer surplus of DP is higher than that of SP at WTP=100, the opposite results are shown when WTP is over 100.

A comparison of provider revenue also shows different results depending on the level of static price.

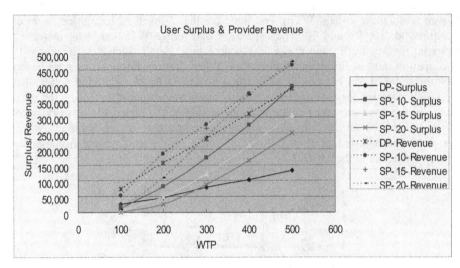

Fig. 3. Comparison of Economic Performance

If we consider QoS-related disutility for consumer surplus, the packet loss in Figure 2 should be applied to decrease consumer surplus. The experiment shows various results according to applying different weights on QoS parameters; the stricter the level of required QoS, the greater the weight of QoS parameter that should be applied.

According to the result of this kind of experiment we can't conclude which policy is better from the viewpoint of consumer surplus. However, we can argue that the consumer surplus of DP shows greater stability and is higher than those of SP when the relative weight of the disutility factor increases.

5   Summary

In this paper, we presented a dynamic pricing mechanism in which socially optimized price is decided according to market demand of network resources. In the Diffserv environment, a service provider can set the bandwidth price for multiple classes in order to maximize his revenue; while a user can decide the amount of resource consumed in order to maximize his own utility.

We have modeled traffic demand from total willingness-to-pay of users and the demand elasticity of price in the market. Our basic assumption was that entering traffic to network can be controlled by price adjustment in which the value is decided from the viewpoint of economic theory.

Our simulation results show that our dynamic pricing model is superior to static pricing for engineering efficiency such as packet loss and link utilization. But economic efficiency is dependent on whether disutility factors, such as loss rate or queuing delay, are considered. If we didn't regard these factors, some static pricing policy would have shown better results than that of dynamic pricing. However, the more significant these factors, the better the economic efficiency of dynamic pricing.

References

1. A.M. Odlyzko, "The Economics of the Internet: Utility, Utilization, Pricing, and Quality of Service," Technical Report. AT&T Research Lab, 1998
2. J.K. MacKie-Mason and H.R. Varian, "Pricing the congestible network resources," IEEE JSAC 13:1141-1149, 1995
3. A. Gupta, D.O. Stahl, and A.B. Whinston, "Pricing of services on the Internet," Technical Report, University of Texas, Austin, 1997
4. S. Shenker, D. Clark, D. Estrin, and S. Herzog, "Pricing in computer networks: Reshaping the Research Agenda," ACM Computer Communication Review, 1996
5. D.D. Clark, "Internet cost allocation and pricing," Internet Economics, L.W. McKnight and J.P. Bailey, Eds, MIT Press 1997
6. S. Kalyanaraman and T. Ravichandran, "Dynamic Capacity Contracting: A Framework for Pricing the Differentiated Services Internet," Int'l Conf. Information and Computation Economics, 1998
7. F.P. Kelly, "Charging and Rate Control for Elastic Traffic," European Transactions on Telecommunications, 8:33-37, 1997
8. F.P. Kelly, A.K. Maullo, and D.K.H. Tan, "Rate Control in Communication Networks: Shadow Prices, Proportional Fairness and Stability," Journal of Operations Research Society, 49:237-252, 1998
9. S.H. Low and D.E. Lapsley, "Optimization Flow Control – I: Basic Algorithm and Convergence," IEEE/ACM Transactions on Networking, 7(6): 861-875, 1999
10. H.R. Varian, Microeconomic Analysis (3rd edition), 1992, New York, Norton
11. C. Courcoubetis and R. Weber, Pricing Communication Networks, 2003, Wiley
12. T. Li, Y. Iraqi and R. Boutaba, "Tariff-based Pricing and Admission Control for Diffserv Networks," INM (International Network Management) 2003
13. M. Baglietto, R. Bolla, F. Davoli, M. Marchese, A. Mainero, and M. Mongelli, "A unified model for a pricing scheme in a heterogeneous environment of QoS-controlled and Best Effort connections," SPECTS 2002
14. X. Wang and H. Schulzrinne, "An Integrated Resource Negotiation, Pricing, and QoS Adaptation Framework for Multimedia Applications," IEEE J. of Selected Areas in Comm. 18(12): 2514-2529, 2000
15. X. Wang and H. Schulzrinne, "Pricing Network Resources for Adaptive Applications in a Differentiated Services Network," IEEE Infocom 2001
16. M. Yuksel and S. Kalyanaraman, "Distributed Dynamic Capacity Contracting: A Congestion Pricing Framework for Diff-Serv," MMNS 2002
17. C. Bouras and A. Sevasti, "A new pricing mechanism for a high-priority Diffserv-based service," ICACT 2004
18. P. Marbach, "Analysis of s Static Pricing Scheme for Priority Services," IEEE/ACM Transactions on networking, Vol. 12, No.2: 312-325, April 2004

Towards an Autonomic Service Architecture

Ramy Farha[1], Myung Sup Kim[1],
Alberto Leon-Garcia[1], and James Won-Ki Hong[2]

[1] Dept. of Electrical and Computer Eng'g, University of Toronto, Toronto, Canada
{ramy.farha,myungsup.kim,alberto.leongarcia}@utoronto.ca
[2] Dept. of Computer Science and Eng'g, POSTECH, Pohang, Korea
jwkhong@postech.ac.kr

Abstract. Traditional telecommunications service providers are undergoing a transition to a shared infrastructure in which multiple services will be offered to customers. These services will be introduced, modified, and retired at a pace that tracks changing requirements and demands. In order to be cost-effective, these services will need to be delivered over a shared infrastructure that is managed to support delivery requirements at a given point in time. In this paper, we present an Autonomic Service Architecture (ASA) for the automated management of networking and computing resources. ASA ensures the delivery of services according to specific agreements between customers and service providers.

1 Introduction

The transition from circuit-based telephone networks to IP-based packet networks presages the replacement of traditional voice telephony service by a broad array of media-rich, personalized and context-aware services that need to provide immediacy, reliability, and consistency of quality levels. Service providers (SPs) will need to be able to deploy, maintain, and retire services quickly according to demand. This future service environment requires a new service delivery framework that can exploit the capabilities of IP networks and provide required service richness, agility, and flexibility. From the information technology (IT) world, autonomic computing [1] is touted as the means to providing a rich set of IT services over a common computing infrastructure. A key feature of autonomic computing is the automated management of computing resources. The application of autonomic management principles to ensure the delivery of telecommunications services is largely unexplored. In this paper, we introduce an Autonomic Service Architecture (ASA) to address this need.

As mentioned before, the first work in the autonomics wave has been the Autonomic Computing proposal by IBM. However, IBM's approach exclusively focuses on computing resources for IT services delivery. Our work aims to expand this basic view to include telecommunications services, which consist of both computing and networking resources. Another proposal called Autonomic Communication [2] has similar aims to IBM's Autonomic Computing, except that it focuses on individual network elements, and studies how the desired

T. Magedanz, E.R.M. Madeira, and P. Dini (Eds.): IPOM 2005, LNCS 3751, pp. 58–67, 2005.
© Springer-Verlag Berlin Heidelberg 2005

element's behavior is learned, influenced or changed, and how it affects other elements. Our work is focused on services, and therefore is a top-down approach as compared to this bottom-up approach. In addition, research projects such as Autonomia [3], AutoMate [4], and Oceano [5] are using the autonomic concept in various ways. Autonomia provides dynamically programmable control and management to support development and deployment of smart applications. AutoMate enables development of autonomic Grid applications that are context-aware, self-configuring, self-composing, and self-optimizing. Oceano is developing a prototype for a scaleable infrastructure to enable multi-enterprise hosting on a virtualized collection of hardware resources. The closest work to ours is that of HP [6], which proposes an architecture and some algorithms for a service-oriented control system that constantly re-evaluates system conditions and re-adjusts service placements and capacities. The control system is organized as an overlay topology with monitoring and actuation interfaces to underlying services. The difference between our work and these approaches is that they consider specific services to which their design is appropriate. We propose a generic architecture which applies to all services.

The main contribution of our paper is that it proposes ASA as a generic architecture for autonomic service delivery by SPs. ASA is service-independent, and defines a resource management model based on virtualization. The rest of this paper is structured as follows. Section 2 describes ASA. Section 3 illustrates the operation of the Autonomic Resource Broker (ARB), the key component of ASA. Section 4 concludes this paper.

2   Autonomic Service Architecture

The players involved in the delivery of a service are the customers and the SPs. We define a service as the engagement of resources for a period of time according to a contractual relationship between the customers and the SPs. Resources are physical and logical components used to construct services. Service Level Agreements (SLAs) are contracts between SPs and customers [7], which are critical in guaranteeing service delivery. Service management ensures that SLAs are met, and that the necessary resources for the service delivery are provided.

After customers and SPs negotiate the SLA, ASA will manage the service in order to meet SLAs without SP's intervention. In this section, we will present ASA according to a layered view, where services are built on underlying virtual and physical resources. If the problems incurred are too complex to be handled autonomically, manual adjustments are needed. When customers purchase a service from a SP, they can themselves offer this service to other customers, becoming SPs to those customers. ASA ensures End-to-End (E2E) service delivery, where SPs negotiate with other SPs involved in the E2E delivery, to which they become customers.

The autonomic service architecture (ASA) is driven by our view that "everything is a service". We identify two types of services, however: Basic services, which cannot be broken down anymore into other services, and which mainly consist of the underlying resources, and Composite services, which are composed from basic services as well as from other composite services.

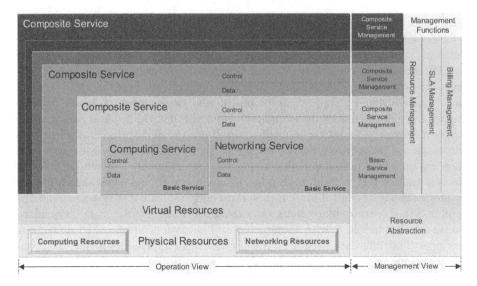

Fig. 1. Autonomic Service Architecture Layered View

Fig. 1 shows ASA's layered view of services. The lowest layer consists of physical resources engaged in the delivery of the service. The middle layer consists of an abstraction of the physical resources into virtual resources according to metrics specifying the characteristics of the physical resources. The upper layers consist of services composed using these underlying resources. Vertically, services are broken into two views: Operation and Management. The Operation view consists of the control and data planes at different layers, while the Management view consists of the management structure needed to manage these services.

2.1 Operation View

The operation view consists of a layering of resources.

Physical Resources Layer. This layer consists of the physical resources that the SP has at its disposal. These resources are either computing (Servers, Workstations, Storage, Clusters), and/or networking (Routers, Switches, Links).

Virtual Resources Layer. This layer abstracts physical resources into virtual resources. Virtualization allows the SP to deal with resources at its disposal to create services *independent* of the actual physical resources. Every service views its needed resources as an array of metrics from a unified language we will define, called the Common Resource Format (CRF). The motivation for CRF is similar to the motivation that drove IBM to propose the Common Base Event (CBE) model [8] to translate proprietary application logs into standard CBE format. From physical to virtual resources, translation is needed from proprietary formats to CRF using an adapter. Virtualization depends on the types of resources involved. Fig. 2 shows three different types of virtual resources:

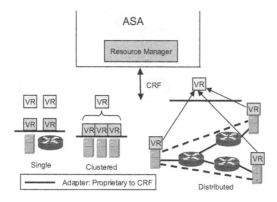

Fig. 2. Virtualization of Resources

1 **Single resources:** Consist of a single physical resource (router, server).
2 **Clustered resources:** Consist of multiple physical resources at a certain geographical location (cluster).
3 **Distributed resources:** Consist of multiple physical resources, geographically dispersed, virtualized to look as an aggregate resource (grid [9]).

Basic Services Layer. Some virtual resources, such as Virtual Networks [10], can be offered directly to customers by SPs as basic networking services, consisting of guaranteed IP transport. In other situations, basic services, which are bought from other SPs according to SLAs, become virtual resources at the disposal of the purchasing SP's composite services.

Composite Services Layer. Composite services consist of several basic services and/or composite services. The composite services can be offered directly to customers. Otherwise, the composite services are virtual resources to other composite services. The process of service composition is hierarchical and recursive, and continues until the composite service is offered to customers.

2.2   Management View

The main task of ASA consists of managing the resources available to the SP in order to meet changes in service demands and user requirements. All management functions (Resource, SLA, Billing) are performed by the Autonomic Resource Broker (ARB), a self-managing entity whose role is to ensure automated delivery of services. A key concept in the IBM autonomic computing architecture is the autonomic element, a component that is responsible for managing its own behavior in accordance with high-level policies, and for interacting with other autonomic elements. In ASA, the Autonomic Resource Brokers (ARBs) are the analogy of autonomic elements. We define the ARB as the component responsible for managing a service instance in accordance with policies, by interacting with underlying resources, and with other ARBs to provide or consume services. ARBs follow policies to ensure SLAs between customers and SPs are met.

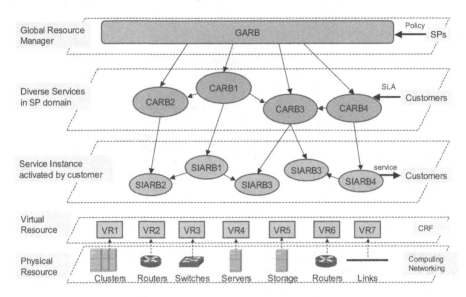

Fig. 3. Hierarchical Management View

When customers activate service instances, these instances are managed by SIARBs (Service Instance ARBs). The multiple service instances of a particular service offered by a SP are managed by CARBs (Composite ARBs). Some CARBs, called basic CARBs, manage the basic services consisting of virtual resources. Other CARBs, called composite CARBs, manage composite services. The different services offered by a SP are managed by a GARB (Global ARB), which handles all the resources available at this SP's disposal. The hierarchical structure of ARBs is shown in Fig. 3. Note that ASA is based on service-oriented architectures [11] for interactions between ARBs, and with underlying resources.

3   Autonomic Resource Broker Architecture

These ARBs are self-managed according to high-level policies. ARBs handle the autonomic operation of ASA. Fig. 4 shows the ARB's internal architecture and the flow of information between the different ARB components.

3.1   Information Bases

Information Bases store the information needed for ASA to autonomically deliver services, in a self-configuring, self-optimizing, self-healing, and self-protecting way. Information Bases can be classified into five logical groupings:

Customer Information Base (CIB): Contains information about customers, such as personal data, list of services subscribed to with their SLA, and bill.

Service Information Base (SIB): Contains information about the service instances activated by the customers, such as parties involved (customer and

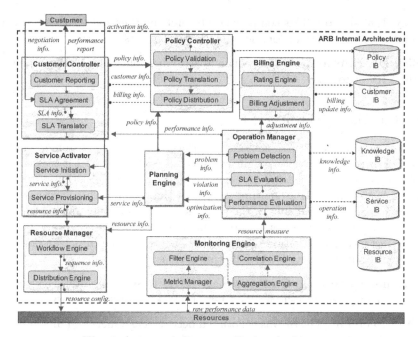

Fig. 4. Autonomic Resource Broker Architecture

SP), SLA agreed upon, types of resources needed, amount of each resource type needed, billing plan, and operation history.

Resource Information Base (RIB): Contains information about the resources available, such as types of resources and quantity.

Policy Information Base (PIB): Contains policies created at runtime, or entered manually. These policies are service-based. There are policies regulating the operation of each ARB component, as well as providing SLA Templates for services offered and specifying types of resources needed for services offered.

Knowledge Information Base (KIB): Contains information for use in case problems arise and remedy actions can be taken based on a previous occurrence of the problem, such as problem description, problem cause, time of occurrence, parties involved, elaborated solutions, and effect of solutions.

3.2 Policy Controller

Policies are created initially by SPs, or at runtime as a result of customers activating services. Existing policies are updated as a result of service demand and load variations. The Policy Controller entails the following actions:

Policy Validation. The creation or update of policies could lead to conflicts, redundancy, inconsistency, and infeasibility. This component ensures no such problems occur and remedies to them.

Policy Translation. This component interprets policies and translates them to an understandable format for use by ARB components.

Policy Distribution. This component distributes policies to ARB components, to the PIB, or to other ARBs.

3.3   Customer Controller

This component constitutes the only interface between the customers and the SPs. This is part of the self-configuring aspect of ASA. Note that it is essential to differentiate between the SLA Agreement and the Service Initiation procedures. The SLA Agreement is a one-time operation that occurs when customers buy the service from SPs. Of course, the customer, using this same procedure, could later modify the service bought, but this operation does not occur for each service instance activation by the customer, as the Service Initiation does. The Customer Controller entails the following actions:

Customer Reporting. Customers have access to some monitoring results, in order to allow them to switch SPs if performance is not satisfactory.

SLA Agreement. The SLA is negotiated between customers and SPs, following a negotiation protocol we define.

SLA Translator. Once the SLA negotiation between the customer and SPs is completed, the SLA Translator creates/updates policies on the fly to regulate the purchased service's delivery. In addition, the SP fills the CIB with customer and billing information. Note that the SLA Translator itself is policy-based, i.e. its operations are regulated by manual policies entered by the SPs which guide its operations, based on services involved, customer types, or other criteria.

3.4   Service Activator

This component is invoked upon service activation by customers. This is part of the self-configuring and self-optimizing aspects of ASA. The Service Activator entails the following actions:

Service Initiation. Customers activate services they already bought. Information, related to the service activated and to the customers, is retrieved and used for Service Provisioning.

Service Provisioning. As mentioned previously, ASA was built based upon the basic premise that "everything is a service". Using this approach to services, composite services are composed out of basic services and of other composite services, and provisioning composite services consists of choosing the appropriate amounts of resources to allocate to each service component. The basic services needed are identified, and the objective function for this service is used to optimize provisioning. The amount of resources needed of each type are calculated, and this information is sent to the Resource Manager.

3.5 Resource Manager

This component allocates/provisions resources available as needed by the service instance that was activated. The appropriate ARBs and/or the underlying resources have to be contacted. This is part of the self-configuring and self-optimizing aspects of ASA. The Resource Manager entails the following actions:

Workflow Engine. The resource allocation process is converted to a workflow of actions executed on underlying resources and/or appropriate ARBs.

Distribution Engine. The actions decided by the Workflow Engine are distributed to underlying resources and/or appropriate ARBs.

3.6 Monitoring Engine

This component monitors the raw performance data sent from underlying resources and/or appropriate ARBs. This is part of the self-healing and self-protecting aspects of ASA. The Monitoring Engine entails the following actions:

Metric Manager. There is a need to quantify raw performance data in a common format (CRF) understandable by ARB components to make decisions.

Filter Engine. The mechanisms to filter unwanted data for all ARBs need to be specified. At the ARB components, the tradeoff is between precision and overhead of measurements. The more precise the results need to be, the more measurements we need to perform.

Aggregation Engine. The measurements received after filtering could be aggregated if a new metric such as a summation, average, maximum, or minimum of the measurements collected is needed.

Correlation Engine. The filtered and aggregated measurements are correlated and complex situations are detected, using techniques such as spatial/temporal correlation, and prediction.

3.7 Operation Manager

This component analyzes ARB operations, and detects any abnormal behavior that results from faults, SLA violation, or sub-optimal performance. This is part of the self-optimizing, self-healing, and self-protecting aspects of ASA. The Operation Manager entails the following actions:

Problem Detection. Faults occur when computing or networking components fail. Overloads occur when the demand on a component exceeds the capacity of that component. Congestions occur when the performance of some components degrades due to excessive load. If problems are detected, the Planning Engine is notified. This is part of the self-healing aspect of ASA.

SLA Evaluation. SLAs are evaluated, and violations detected are sent to the Planning Engine, and to the Billing Engine for proper adjustments to the customer bill in the CIB. This is part of the self-optimizing aspect of ASA.

Performance Evaluation. When the operation is satisfactory (no problems or violations), ARB ensures resources are optimally allocated, and if not, notifies the Planning Engine. This is part of the self-optimizing aspect of ASA.

3.8   Planning Engine

This component is considered to be the brain of the ARB. This is part of the self-optimizing, self-protecting, and self-healing aspects of ASA. The inputs to the Planning Engine are:

- Customer entry, i.e. the customer information related to the services where problems have occurred, and their SLAs. These can be obtained from the Customer Information Base (CIB).
- Service entry, i.e. performance requirements for services (SLAs). These can be obtained from the Service Information Base (SIB).
- Policy entry, i.e. policies that restrict allocation of resources and constrain solutions. These can be obtained from the Policy Information Base (PIB).
- Resource entry, i.e. resources at the SP's disposal extracted from the resource pool. These can be obtained from the Resource Information Base (RIB).
- Knowledge entry, i.e. previous comparable situations, where the advocated solutions could be used instead of elaborating new ones. These can be obtained from the Knowledge Information Base (KIB).
- Notifications from the Operation Manager component to indicate problems detected, SLA violations, and sub-optimal performance.

The outputs that can be generated by the Planning Engine are:

- Changes to the Service Provisioning, for instance resources needed to meet service requirements, re-allocation plans to improve service performance.
- Changes to the Resource Manager when the problems are not drastic enough to require re-provisioning of the service instance.
- Changes to the policies regulating the operation of ARB components.

3.9   Billing Engine

This component bills customers using three charges: A flat charge, a usage charge, and a content charge. This allows more flexibility in using the appropriate billing solution appropriate for each service, class of service, or even individual customer. The pricing is adjusted based on congestion and on billing policies to reflect SLA violations, and to allow pricing to be used as a congestion control technique by increasing the prices of the resources when the network is congested. Note that the exact details of the billing plan are determined when the customer buys the service. This is part of the self-healing aspect of ASA.

4 Conclusion

The journey to a fully autonomic service architecture is still in its early stages. This paper illustrates our proposed generic approach towards this goal, using ASA. ASA allows service providers to reduce service delivery costs to customers. ASA is based on two main concepts: virtualization of physical resources using a common language (CRF), and autonomic service delivery using a hierarchy of Autonomic Resource Brokers (ARBs). The hierarchical service view allows ASA to easily expand to next-generation services by allowing flexible, scalable, and recursive service management. ASA is still a conceptual architecture, but its realization for real services is underway. First, we are defining XML formats for the information bases, for policies, for service and SLA templates, as well as elaborating CRF. Second, we are defining interfaces among ARBs, exploring several ARB topologies (Peer-to-Peer, Hierarchical, Hybrid) and assessing them. Third, we are developing algorithms for each ARB functional block (e.g. Service Activator, Operation Manager, Planning Engine etc.). Finally, we are implementing ASA in our Network Architecture Laboratory at the University of Toronto, for specific services such as Voice over IP (VoIP).

References

1. IBM Corporation: An architectural blueprint for autonomic computing. White Paper, (2003)
2. Autonomic Communication: http://www.autonomic-communication.org
3. Xiangdong, D. et. al.: Autonomia: an autonomic computing environment. Proceedings of the IEEE International Performance, Computing, and Communications Conference (2003) 61–68
4. Agarwal, M. et. al.: AutoMate: enabling autonomic applications on the grid. Autonomic Computing Workshop (2003) 48–57
5. Appleby, K. et. al.: Oceano: SLA based management of a computing utility. Proceedings of the IEEE/IFIP International Symposium on Integrated Network Management Proceedings (2001) 855–868
6. Andrzejak, A., Graupner, S., Kotov, V., Trinks, H.: Adaptive Control Overlay for Service Management. Workshop on the Design of Self-Managing Systems, International Conference on Dependable Systems and Networks (DSN) (2003)
7. Leon-Garcia, A., Widjaja, I.: Communication Networks. Mc Graw Hill (2004)
8. Bridgewater, D.: Standardize messages with the Common Base Event model. IBM DeveloperWorks (2004)
9. Talia, D.: The Open Grid Services Architecture: where the grid meets the Web. IEEE Internet Computing Magazine (2002) 67–71
10. Leon-Garcia, A. L. Mason, L.: Virtual Network Resource Management for Next-Generation Networks. IEEE Communications Magazine (2003) 102–109
11. Kreger, H.: Web Services Conceptual Architecture. White Paper, IBM Software Group (2001)

Random Feedbacks for Selfish Nodes Detection in Mobile Ad Hoc Networks

Djamel Djenouri[1], Nabil Ouali[2], Ahmed Mahmoudi[2], and Nadjib Badache[2]

[1] Basic Software Laboratory, CERIST Center of research, Algiers, Algeria
`ddjenouri@mail.cerist.dz`
[2] LSI, USTHB University, Algiers, Algeria
`nabilouali@hotmail.com`, `mahmoudi_pfe@yahoo.fr`, `badache@lsi-usthb.dz`

Abstract. A mobile ad hoc network (MANET) is a temporary infrastructureless network, formed by a set of mobile hosts that dynamically establish their own network *on the fly* without relying on any central administration. Mobile hosts used in MANET have to ensure the services ensured by the powerful fixed infrastructure in traditional networks, the packet forwarding is one of these services.

Resource limitation of MANET's nodes, particulary in energy supply, along with the multi-hop nature of these networks may cause a new problem that does not exist in traditional networks. To save its energy a node may behave **selfishly (no-cooperatively)**, thus it misbehaves by not forwarding packets originated from other nodes, while using their resources to forward its own packets to remote recipients. Such a behavior hugely threatens the QoS (Quality of Service), and particulary the packet forwarding service availability. Some solutions for selfish nodes detection have been recently proposed, but almost all these solutions rely on the monitoring in the promiscuous mode technique of the watchdog [1], which suffers from many problems especially when using the power control technique. In this paper we propose a new approach to detect selfish nodes unwilling to participate in packet forwarding, that mitigates some watchdog's problems. We also assess the performance of our solution by simulation.

Keywords: mobile ad-hoc networks, security, selfishness, packet forwarding, GloMoSim.

1   Introduction

In some MANETs applications, such as in battlefield or rescue operations, all the nodes have a common goal and their applications belong to a single authority, thus they are *cooperative by nature*. However, in many civilian applications, such as networks of cars and provision of communication facilities in remote areas, nodes typically do not belong to a single authority and they do not pursue a common goal. In such self-organized networks forwarding packets for other nodes is not in the direct interest of any node, so there is no good reason to trust nodes and assume that they always cooperate. Indeed, each node tries to

T. Magedanz, E.R.M. Madeira, and P. Dini (Eds.): IPOM 2005, LNCS 3751, pp. 68–75, 2005.

save its resources, particularly its battery power which is a precious resource. Recent studies show that most of the nodes energy in MANET is likely to be devoted to forward packets in behalf of other nodes. For instance, Buttyan and Hubaux simulation studies [2] show that; when the average number of hops from a source to a destination is around 5 then almost 80% of the transmission energy will be devoted to packet forwarding.

Therefore, to save energy nodes may misbehave and tend to be *selfish*. A selfish node regarding the packet forwarding process is a node which takes advantage of the forwarding service and asks others to forward its own packets, but does not actually participate in this service. Some solutions have been Recently proposed. Almost all these solutions, however, rely on the watchdog [1] technique which suffers from many problems. It might cause false accusation, especially when using the power control technique employed by some new power-aware routing protocols following the watchdog's proposal.

Our purpose in this work is to propose a novel solution to mitigate some of these problems.

The remainder of this paper is organized as follows: In the next section we present related work, and then we present and discuss our solution in section 3. Section 4 is devoted to the simulation-based performance evaluation. Finally, section 5 concludes the paper and summarizes our perspectives.

2   Related Work

To the best of our knowledge, Marti et al. [1] are the first who dealt with the problem of nodes misbehavior on packet forwarding, they proposed the *watchdog* which they implemented with the dynamic source routing protocol (DSR) [3]. The watchdog lies on monitoring neighbors in the promiscuous mode, i.e each node in the source route monitors its successor after it sends it a packet to forward by overhearing the channel and checking whether the monitored relays the packet. The monitor accuses a monitored node as misbehaving when it detects that this latter drops more than a given number (threshold) of packets. This basic technique have been used by almost all the subsequent solutions. Nevertheless, it suffers from some problems, especially when using the power control technique, employed by some new power-aware routing protocols following the watchdog's proposal [4] [5] [6].

Assume three aligned nodes: A, B and C, such that A sends B a packet and monitors its forwarding to C, and lets assume that B uses the power control technique. When A is closer to B than C, B could circumvent the watchdog by using a transmission power strong enough to reach A, but less than the one required to reach C, which is power efficient for B. On the other hand, when C is closer to B than A, and B behaves correctly but uses the power control technique, A could not overhear B's forwarding to C, which might results in false detections when the number of packets falsely detected exceeds the configured threshold. Further, packet collisions either at C or A, during the monitoring, could cause problems. When B's forwarding causes a collision at C, the former could circumvent to A by not retransmitting the packet. On the other side, if B's forwarding results in a collision at A, A could falsely note a B's packet dropping.

In [7], Yang et al. describe a unified network layer solution to protect both routing and data forwarding in the context of AODV. Michiardi and Molva [8] suggest a generic reputation-based mechanism, namely CORE (COllaborative REputation Mechanism to enforce node cooperation in MANETs), that can be easily integrated with any network function. CONFIDANT (Cooperation Of Nodes Fairness in Dynamic Ad hoc Networks) is another interesting reputation-based solution, proposed by Buchegger and Le Boudec [9] [10]. It relies on DSR [3] used as benchmark in the GloMosim-based simulation study performed by the authors to evaluate the new DSR fortified by CONFIDANT.

All these solutions, however, rely on the watchdog technique in their monitor component.

Buttyan and Hubaux [11] propose an efficient preventive economic-based approach stimulating nodes to cooperate, which is modeled and analyzed in [2]. The authors introduce what they called *virtual currency* or *nuglets*, along with mechanisms for charging/rewarding service usage/provision. The main idea of this technique is that nodes which use a service must pay for it (in nuglets) to nodes that provide the service. Other stimulating preventive approaches are based on game theory, such as [12].

These preventive solutions motivate nodes to cooperate, but do not aim at detecting the misbehaving nodes contrary to the previous solutions.

In [13], Papadimitratos and Haas present the SMTP protocol. It is a hybrid solution that mitigates the selfishness effects (packets lost) by dispersing packets, and detects the selfish misbehavior by employing end-to-end feedbacks. This kind of feedbacks allows the detection of the routes containing selfish nodes, but fails to detect these nodes. To overcome this problem, Kargl et al. [14] propose *iterative probing*, that detects links containing selfish nodes, but fails to detect appropriate nodes. To find the appropriate node on a link after an iterative probing, authors propose what they called *unambiguous probing*, which is based on the watchdog, thus suffers from its problems.

3   Novel Solution

3.1   Solution Overview

We define a new kind of feedbacks we call *two-hop ACK*, an ACK that travels two hops. Node C acknowledges packets sent from A by sending this latter via B a two-hop ACK.

Node B could, however, escape from the monitoring without being detected by sending A a *falsified* two-hop ACK. Note that performing in this way is power economic for B, since sending a short packet like an ACK consumes too less energy than sending a data packet. To avoid this vulnerability we use an asymmetric cryptography based strategy as follows:

Node A generates a random number and encrypts it with C's public key (PK) then appends it in the packet's header as well as A's address. When C receives the packet it gets the number back, decrypts it using its secret key (SK), encrypts it using A's PK, and puts it in a two-hop ACK which is sent back to A via B. When A receives the ACK it decrypts the random number and checks if the number within the packet matches with the one it has generated, to validate B's forwarding regarding the packet in question. However, if B does not forward the packet A will not receive the two-hop ACK, and it will be able to detect this dropping after a time out. A will then accuse B as selfish when the number of packets dropped it detects exceeds a given threshold.

This encryption strategy needs a security association between each pair of nodes to ensure that nodes share their PK with each other. This requires a key distribution mechanisms which is out of the scope of this paper, but a mechanism like [15] or [16] can be used.

The watchdog's problems are mitigated with this approach, since B's forwarding validation at A is not only related to B's transmission, but to C's reception. Still, the problem with this first solution is that it requires a two-hop ACK for each packet, which might result in important overhead. To decrease this cost, we propose to *randomize* the ACK request. viz. A does not ask C an ACK for each packet, but when sending a packet to forward, it *randomly* decides whether it asks an ACK or not, with a probability p (probability of requesting an ACK). This *random* selection strategy prevents the monitored node from deducting which packets contain ACK requests. Note that getting such information allows a selfish to drop packets with no requests without being detected.

The probability p is either continuously decreased (resp increased) with α after each ACK request during a series of ACK requests[1] (resp sending a packet without requesting an ACK during a series of no-requests) till reaching 0 (resp 1), or switched after a series of requesting (res no-requests) to a non-request (res request). In these two latter cases of switching, p takes the value θ, the initial probability, which is continually updated as follows:

It is set to 1 upon a lack of a requested ACK (after the timeout), and decreased each time the requested ACK is received, till reaching the minimum value θ_0. This way, more trust is given to well-behaving nodes, and by setting θ to 1 the ACK request is enforced after a lack of ACK.

3.2 Discussions

Unlike the current detective solutions that are based on the promiscuous mode monitoring (the watchdog), ours relays on a new technique, namely the two-hop ACK. The monitoring (node A) validates the monitored (node B) forwarding when it receives an ACK from the successor of this latter (node C). This process can be generalized along the path for each consequent two hops until the destination, and efficient encryption/decryption operations have been added in order to authenticate the two-hop ACKs and secure the solution against spoofing attacks.

Getting rid of the promiscuous mode makes our solution independent of transmission powers, and resolves the watchdog problems related to the employment of the power-control technique. Further, it resolves the receiver collision

---

[1] A series of ACK requests is a series of packets, for which A asks C ACKs

problem presented previously. When a collision appears at C, B should retransmit the packet, otherwise A will not validate its forwarding. This because B's forwarding will not be validated at A until C really receives the packet and sends back the two-hop ACK, unlike the watchdog where the validation is only related to B's first transmission.

Moreover, The two-hop ACK technique allows to detect the appropriate selfish node, unlike the end-to-end ACKs [13] and the iterative probing [14].

As illustrated, authentication of the two-hop ACK packet is ensured by employing encreption/decreption operations on a random number, generated by the monitor and piggybacked to the monitored packet. These operations have minor impact on computation overhead, since they are applied merely on the random number and not on the whole packet holding it. Note that we avoided the use of digital signatures in order to avoid useless packet's hash computation.

Instead of requiring a two-hop ACK for each packet, we proposed to *randomize* the ACK request, where the monitor *randomly* decides whether it asks an ACK or not, with a certain probability which is continuously updated in such a way to give more trust (less ACK requirements) to well-behaving nodes, and enforce requirements after an ACK reception failure. This *random* selection strategy prevents the monitored node from deducting which packets contain ACK requests, thus from dropping two-hop ACK free packets[2].

Still, like the watchdog and all the currently proposed solutions, the challenging problem of cooperative misbehavior remains untreated with our proposal. That is, if B and C collude, such as B does not correctly monitor C (i.e it does not report back when C drops packets), C will not be detected. This problem represents an open research topic.

4   Simulation Study

To asses the proposed protocol performance we have driven a GloMoSim-based [17] simulation study, that we will present hereafter.

We have simulated a network of 50 nodes, located in an area of $1500 \times 1000 m^2$, where they move following the random way-point [18] model with an average speed of 1m/s, for 900 seconds (the simulation time). To generate traffic, we used three CBR sessions between three pairs of remote nodes, each consists of continually sending a 512 bytes data packet each second. On each hop, each data packet is transmitted using a controlled power, according to the distance between the transmitter and the receiver.

We compare two versions of our protocol, 2HopACK and Random 2HopACK, as well as the watchdog (WD), with regard to the selfish detection rate, the false detection rate (rate of false accusations as selfish) and the number of two-hops ACKs (which represents the overhead). We measured these metrics vs the selfish nodes rate, which represents the rate of nodes that behave selfishly and drop packets they are asked to relay. Each point of the plots presented hereafter has been obtained by averaging five measurements with different seeds. Note that

---

[2] Packets that do not require a two-hop ACK

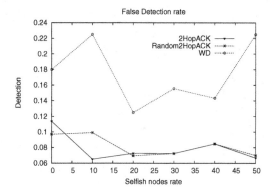

Fig. 1. False detection vs. selfish rate

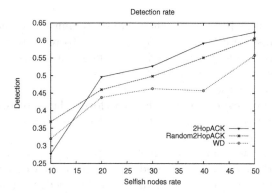

Fig. 2. True detection vs. selfish rate

we implemented our protocol with DSR for this simulation, like the watchdog. However, it can be implemented with any source routing protocol. Also note that WD requires no kind of ACK, so the last metric (number of two-hop ACK) concerns merely our protocol's versions.

The first version of our protocol requires an ACK for each packet, while the second one uses the efficient technique of randomizing the ACK request, which reduces the overhead especially when the selfish nodes rate is low, as shown in figure 1. However, the cost of this overhead decreasing is a minor loss in detection efficiency, as shown in figure 2. But we can clearly see in the same figure that both versions have better detection than WD. Figure 3 illustrates how our protocol (the two versions) decreases hugely the false detection rate compared with WD.

5 Conclusion and Future Work

In this work we have focused on the selfish nodes detection problem, and we proposed an approach that mitigates some watchdog's drawbacks. Our solution is based on the two-hop ACK, it allows to detect the selfish node, unlike the

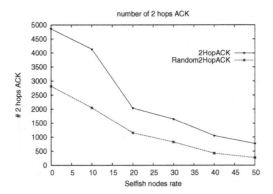

Fig. 3. Number of two-hop ACK vs. selfish rate

end to end ACK that just detects the route containing such a node, or iterative probing which just detects the appropriate link. To reduce the overhead while keeping efficiency, we have suggested to ask two-hop ACKs at random points, instead of of asking them for each data packets.

We have assessed the performance of our solution by a simulation study, whose results show how our approach outperforms the watchdog, especially regarding false detections. The simulation results also show how the random requesting strategy reduces the overhead.

As perspective, we plan to complete the proposal by defining actions that have to be taken when a node is accused as a selfish, and particulary by proposing a mechanism allowing nodes to exchange their knowledge regarding nodes that behave selfishly.

References

1. Marti, S., Giuli, T., Lai, K., Baker, M.: Mitigating routing misbehavior in mobile ad hoc networks. In: ACM Mobile Computing and Networking, MOBICOM 2000. (2000) 255–65
2. Buttyan, L., Hubaux, J.P.: Stimulating cooperation in self-organizing mobile ad hoc networks. ACM/Kluwer Mobile Networks and Applications, Vol 8, N 5 (2003)
3. David, B., David, A.: Dynamic source routing in ad hoc wireless networks. Mobile Computing, Chapter 5 (1996) 153–181
4. Djenouri, D., Badache, N.: New power-aware routing for mobile ad hoc networks. Accepted in the International Journal of Ad Hoc and Ubiquitous Computing (Inderscience) (2005 (to appear))
5. Doshi, S., Brown, T.: Minimum energy routing schemes for a wireless ad hoc network. In: IEEE INFOCOM 2002. (2002)
6. Djenouri, D., Badache, N.: Simulation performance evaluation of an energy efficient routing protocol for mobile ad hoc networks. In: IEEE International Conference on Pervasive Services (ICPS'04), American University of Beirut (AUB), Lebanon (2004)

7. Yang, H., Meng, X., Lu, S.: Self-organized network layer security in mobile ad hoc networks. In: ACM MOBICOM Wireless Security Workshop (WiSe'02), Georgia, Atlanta, USA. (2002)
8. Michiardi, P., Molva, R.: Core: A collaborative reputation mechanism to enforce node cooperation in mobile ad hoc networks. In: Communication and Multimedia Security 2002 Conference, Portoroz, Slovenia. (2002)
9. Buchegger, S., Boudec, J.Y.L.: Performance analysis of the confidant, protocol cooperation of nodes fairness in dynamic ad hoc networks. In: Third ACM International Symposium on Mobile Ad Hoc Networking and Computing (MobiHoc'02), Lausanne, Switzerland. (2002) 80–91
10. Buchegger, S., Le-Boudec, J.Y.: A robust reputation system for p2p and mobile ad-hoc networks. In: Second Workshop on the Economics of Peer-to-Peer Systems. (2004)
11. Buttyan, L., Hubaux, J.P.: Nuglets: a virtual currency to stimulate cooperation in self-organized mobile ad hoc networks. Technical report No. DSC/2001/001, Swiss Federal Institution of Technology, Lausanne, Switzerland (2001)
12. Srinivasan, V., Nuggehalli, P., F.Chiasserini, C., R.Rao, R.: Cooperation in wireless ad hoc networks. In: IEEE INFOCOM'03, San Francisco, California, USA. (2003)
13. Papadimitratos, P., Haas, Z.J.: Secure data transmission in mobile ad hoc networks. In: ACM MOBICOM Wireless Security Workshop (WiSe'03), San Diego, California, USA. (2003)
14. Kargl, F., Klenk, A., Weber, M., Schlott, S.: Advanced detection of selfish or malicious nodes in ad hoc networks. In: 1st European Workshop on Security in Ad-Hoc and Sensor Networks, ESAS 2004. (2004)
15. Capkun, S., Buttyan, L., Hubaux, J.P.: Self-organized public-key management for mobile ad hoc networks. IEEE Transactions on Mobile Computing, Vol.2, No.1 (2003) 52–64
16. Weimerskirch, A., Westhoff, D.: Zero common-knowledge authentication for pervasive networks. In: Selected Areas in Cryptography. (2003) 73–87
17. Zeng, X., Bagrodia, R., Gerla, M.: Glomosim: A library for the parallel simulation of large-scale wireless networks. In: proceeding of the 12th Workshop on Parallel and distributed Simulation. PADS'98. (1998)
18. Djenouri, D., Derhab, A., Badache, N.: Ad hoc networks routing protocols and mobility. International Arab jornal of Information Technology (2005 (to appear))

A Packet Class-Based Scheme
for Providing Throughput Guarantees to TCP Flows

Lluís Fàbrega, Teodor Jové, Pere Vilà, and José Marzo

Institute of Informatics and Applications (IIiA), University of Girona,
Campus Montilivi, 17071 Girona, Spain
{fabrega,teo,perev,marzo}@eia.udg.es

Abstract. TCP flows generated by applications such as the web or ftp require a minimum network throughput to satisfy users. To build this service, we propose a scheme with Admission Control (AC) using a small set of packet classes in a core-stateless network. At the ingress each flow packet is marked as one of the set of classes, and within the network, each class is assigned a different discarding priority. The AC method is based on edge-to-edge per-flow measurements, and it requires flows to be sent at a minimum rate. The scheme is able to provide different throughput to different flows and protection against non-responsive sources. We evaluate the scheme through simulation in several network topologies with different traffic loads consisting of TCP flows that carry files of varying sizes. In the simulation, TCP uses a new algorithm to keep the short-term sending rate above a minimum value. The results prove that the scheme guarantees the throughput to accepted flows and achieves high utilization of resources, similar to the ideal results of a classical hop-by-hop AC.

1 Introduction

Internet traffic is currently dominated by TCP connections carrying files generated by applications such as the web, ftp, or peer-to-peer file sharing [1]. The users of these applications expect no errors in the file transfer and the best possible response time. The source breaks up the file and sends a flow of packets at a certain sending rate, which the network delivers to the destination with variable delays and losses. Losses are detected and recovered by TCP through retransmission, which adds more delay (increasing the transfer time) and may also cause duplicated packets (which are discarded by the destination). From the point of view of the network, the decisive Quality of Service (QoS) parameter is the average receiving rate or network throughput (which includes the duplicates).

A basic feature of TCP flows is their elastic nature. Sources vary the sending rate (up to the capacity of the output link) to match the maximum available throughput in the network path. Since the available throughput change over time, TCP uses rate-adaptive algorithms that increase and decrease the sending rate in order to match these variations and minimize packet loss. TCP increases the sending rate if packets are correctly delivered and decreases it if packets are lost [2]. On the other hand, another feature of TCP flows is the heavy tail behavior of the file size distribution observed in traffic measurements [1], which means that most elastic flows carry short files and a few flows carry very long files.

Although the traditional view is that elastic flows do not require a minimum throughput, unsatisfied users or high layer protocols impose a limit on the file transfer

T. Magedanz, E.R.M. Madeira, and P. Dini (Eds.): IPOM 2005, LNCS 3751, pp. 76–87, 2005.

time. This situation implies a waste of resources, which can get even worse if the transfer is tried again [3]. Moreover, in commercial Internet services, users will pay extra for the performance they require. Hence, there is a minimum throughput required or desired by users.

Elastic flows are satisfactorily supported by a guaranteed minimum throughput service, which provides a minimum throughput and, if possible, an extra throughput. There is an input traffic profile, based on some sending traffic parameters (average rate, burst size, etc.), which defines the desired minimum throughput. If the average sending rate is within the profile, packets have a guaranteed (minimum) delivery; otherwise packets have a possible (extra) delivery only if there are available resources (which are shared among flows according to a best-effort or other policy).

The service delivery is defined in a bilateral agreement (Service Level Agreements or SLAs) between the provider and user (an end-user or a neighboring domain). The agreement specifies that the user can ask for the service for each flow of an aggregation of any number of flows to any destination, and also for any flow's minimum throughput, as long as a contracted value is not exceeded. The agreement also specifies the minimum percentage of requests the provider should satisfy (which is usually expected to be high).

Throughput services have been proposed in the past. The traditional network service is the best-effort service, which together with TCP rate-adaptive algorithms [2], have the goal of providing a fair throughput service. Another proposal is the Assured Service [5], defined in the Differentiated Services (Diffserv) architecture [6]. It is a proportional throughput service able to provide different throughputs to different users. However, in congestion (when users' demands exceed resources), both services cannot provide a desired minimum throughput to any flow. Congestion could be avoided using resource overprovisioning, but this is a highly wasteful solution. If more efficient provisioning is used congestion can be dealt with by Admission Control (AC). Its goal is to provide a desired minimum throughput to the maximum possible number of flows in congestion.

A classical distributed AC method with hop-by-hop decision, based on per-flow state and signaling in the core, is not appropriate for elastic flows because of the low scalability and the high overhead of explicit signaling. Therefore other kind of AC methods have been proposed [7][8][9]. They have in common that avoid the use of per-flow signaling, reduce the per-flow state to the edge or do not need it, and use measurements.

We propose a scheme for a guaranteed minimum throughput service using Admission Control (AC), with simple and scalable mechanisms. The scheme does not need per-flow processing in the core since it only uses a small set of packet classes. At the network ingress each flow packet is marked as one of the classes, and within the network, each class is assigned a different discarding priority. The AC method is based on edge-to-edge, per-flow throughput measurements using the first packets of the flow, and it requires flows to be sent at a minimum rate. In this paper we develop this scheme, initially outlined in [4], and we present simulation results using statistical traffic variations in different network topologies to prove its validity.

The paper is organized as follows. In section 2 we describe our scheme, in section 3 we present simulation results and finally we summarize the paper in section 4.

2 The Proposed Scheme

Our proposal is similar to [9] in that a flow is defined as a sequence of related packets (from a single file transfer) within a TCP connection, a network path is considered and per-flow state is only maintained at the edge. However, our scheme is able to provide different throughput to different flows as well as protection against non-responsive sources. It also uses measurements, but it is built from the Assured Service scheme [5] and the set of AC methods that uses end-to-end per-flow measurements [10][11]. Our purpose is to add AC to the Assured Service scheme also using a small set of packet classes.

Our scheme considers the following assumptions. Firstly, to avoid the use of explicit flow signaling, a list of active flows is maintained at the edge using an implicit mechanism to detect the start and end of flows [9]. Secondly, we assume a network architecture that uses packet classes, such as Diffserv [6]. Thirdly, we assume that there are pre-established logical paths from ingress points to egress points, and each new arriving flow at the edge is assigned to one of them. An example of this mechanism is the Label Switched Paths of MPLS networks [12] when used without a long-term resource reservation. Instead, once a flow is accepted by our AC, resources in the logical path are reserved for the lifetime of the flow. We use the logical paths to pin each flow to a route, so that all flow packets follow the same path where the reservation has been made. Finally, to avoid again the use of explicit flow signaling, we assume that in the agreements each user specifies the desired minimum throughput of flows, e.g., the same value for all flows, according to the application type (ftp, web…) or other criteria.

2.1 The Architecture of the Scheme

When the first packet of a new flow arrives to the network, the flow is assigned to a logical path and the list of active flows is updated at the ingress. The AC evaluates if its minimum throughput requirement (from the corresponding agreement) can be provided without losing the minimum throughput guaranteed to the accepted flows. If the flow is accepted, it receives a guaranteed minimum throughput service; otherwise, it receives a best-effort service.

The kind of AC we propose belongs to the methods based on end-to-end per-flow measurements [10][11], where end-points send a probing flow and measure its experienced QoS to estimate if there are enough available resources in a path to satisfy the flow's request. After this initial time period, called the AC phase, the AC decision is made. In our method [4], the first packets of the flow act as the special probing flow to test the network. The throughput is measured at the egress and then it is sent to the ingress, where it is compared with the requested minimum throughput to make the AC decision.

The whole scheme, including the AC method, uses five packet classes, each one with a different discarding priority. The classes are two A (Acceptance) classes, A_{IN} and A_{OUT}, two R (Requirement) classes, R_{IN} and R_{OUT}, and the BE (Best-Effort) class. The scheme is the following (Fig. 1):

- During the AC phase, at the ingress, packets are marked as R by a traffic meter and a marker: each flow packet is classified as "in" or "out", depending on whether the (measured) average sending rate is smaller or greater than the desired minimum throughput, and then in packets are marked as R_{IN} and out packets as R_{OUT}.
- During the AC phase, at the egress, the throughput of the flow is measured.
- At the end of the AC phase, at the ingress, the measured throughput is compared to the requested minimum throughput: if it is smaller the flow is not accepted; otherwise it is accepted.
- After the AC phase, at the ingress, if the flow has been accepted, its packets are marked as A (A_{IN} or A_{OUT}, depending on the comparison between the average sending rate and the desired minimum throughput), and if it has been rejected, its packets are marked as BE.
- In the output links of routers there is a single FIFO queue (to maintain packet ordering) with discarding priorities for each class in the order (from low to high) A_{IN}, R_{IN}, A_{OUT}, R_{OUT} and BE.

The priorities are chosen so that accepted flows are protected against flows that are in the AC phase. Since the lowest discarding priority class receives the resources in the first place, the next class the remaining ones, and so on, this roughly means that the R packets must be discarded before the A packets (A < R). Specifically, the scheme needs flows in the AC phase to be able to get a measurement of the unreserved throughput in the path ($A_{IN} < R_{IN} < A_{OUT}$). Moreover the remaining extra throughput is given to accepted flows rather than to flows that are in the AC phase ($A_{OUT} < R_{OUT}$).

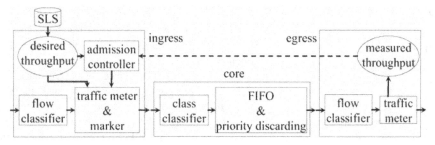

Fig. 1. Functional block diagram of our scheme

2.2 The Measurement Process

The flow's throughput is measured at the egress from the received data packets. A list of active flows that are being measured is maintained at the egress, that is, when a new flow arrives it is added to the list and when the measurement is finished it is removed. The measured throughput is simply the ratio between the total bytes of the received packets during the measurement period divided by T, the duration of this measurement period. Then signaling packets carry the throughput measurement to the corresponding ingress. Note that the AC phase duration is equal to the measurement duration T plus the packet round trip time.

It is important to decide on the value of T. If it is too short the measurement could be incomplete, but if it is too long the method could tend to reject too many flows in some situations. We study the influence of the measurement duration in section 3.

2.3 "Occupancy" Based Reservations

The reservation of resources for an accepted flow is setup by its "in" packets (marked as R_{IN} during the AC phase) when the flow starts to transmit, maintained as long as the flow transmits, and then released when it stops, because the measurements will reflect all these situations. We say that per-flow reservations are based on "occupancy" instead of "state". Per-flow state is only maintained at the edge and per-flow processing in the core is not needed. However, in order for the AC to work properly, sources should never send less than the minimum throughput. Otherwise, if an accepted flow does not use the guaranteed throughput, other, newly arriving flows would be erroneously accepted, and the throughput allocated to them later on might be incorrect.

When TCP transfers a file and there is a minimum available network throughput, the average sending rate during the flow's lifetime is kept approximately above this minimum throughput. However, the short-term fluctuations of the sending rate of a "standard" TCP source might cause some inaccuracies that could reduce the performance of the scheme. In the simulations in section 3 we use a modified TCP source that keeps the short-term sending rate above a minimum value.

3   Evaluation of the Scheme

We have added the functional blocks of our scheme (Fig. 1) to the Diffserv module of the ns simulator [13]. All the functional blocks have been designed for the traffic sent by the modified TCP source. The sending rate meter uses the token bucket algorithm, which is characterized by two parameters, the rate in bps and the burst size in bytes (which constitute the service's input traffic profile). The marker uses five marks that identify the packet classes (section 2.1). The throughput meter starts to measure when the first packet of the flow arrives, then it simply counts the total received bytes during the time of measurement, and finally it notifies the throughput measurement to the ingress router (with the corresponding delay). The FIFO queue with priority discarding for the packet classes is based on the drop-tail algorithm, that is, when the queue is full and a new packet arrives, a packet is discarded (the one with the highest discarding priority).

3.1   The Modified TCP source

We have modified the TCP source of the simulator to keep the short-term sending rate above a minimum value. Instead of sending a burst of packets (the ones that would be allowed by the window) within approximately one round trip time, packets are sent smoothly throughout the same time period. That is, one packet is sent every Δt time units, which is increased and decreased by an algorithm causing the rate variations. The time Δt has a maximum value that guarantees the minimum rate.

Variations of Δt follow proportionally the window variations of the TCP New-Reno algorithms implemented in the ns simulator. Moreover, packet retransmission from the last acknowledged packet is triggered by the usual circumstances (when the corresponding ACK packet is not received during a timeout or when three duplicated ACKs of a previous packet are received [2]), and also when the packet sequence number reaches the end value and the corresponding ACK has not been received yet.

Finally, another feature we have added to the TCP sources takes into consideration the users' impatience, that is, the source stops sending if the file transfer time is too high.

3.2 Traffic Model and Network Topologies

We generate TCP flows that carry a single file from an ingress point to an egress point through a logical path. Each flow is characterized by the file size and the starting time. File sizes are obtained from a Pareto distribution because it approximates reasonably well the heavy tail behavior of the file size distribution observed in measurements [1]. In all simulations the Pareto constant (or tail parameter) is 1.1 and the minimum file size is 10 packets (the packet length is 1000 bytes). With these parameters the distribution has an infinite variance. After generating the values (about 10,000), more than 50% are below 20 packets, the mean σ is about 74 packets and the maximum may even reach 19637 packets. On the other hand, the starting times are obtained from a Poisson arrival process, which is a simple and useful model that has proved to be valid for more realistic scenarios in [14]. This process is characterized by the parameter λ flow/s, i.e. the average number of arrivals per second. Using all these statistical distributions, the average offered traffic load from an ingress point to an egress point through a logical path is equal to $\lambda\sigma$ bps.

We use three network topologies (Fig. 2) with different number of hops and logical paths. In topology 1 there are two logical paths, e0-e3 and e2-e3, in topology 2, three logical paths, e0-e8, e1-e5 and e3-e7, and in topology 3, six logical paths, e0-e9, e2-e9, e3-e9, e5-e9, e6-e9, and e8-e9. The value of the average offered traffic is the same for each path, and it varies from 0.05 to 3 Mbps in steps of 0.3 Mbps.

For all TCP flows we use the same values of the token bucket algorithm, a rate of 90 Kbps and a burst size of two packets, which represent the desired minimum throughput. The user's impatience is twice the desired file transfer time, which means getting a throughput of approximately 45 Kbps.

3.3 Performance Parameters

The goal of the AC method is to provide the desired minimum throughput to the maximum possible number of flows, and therefore we evaluate the utilization of resources as well as the throughput obtained by flows. Firstly, we calculate the throughput of each flow as the ratio between the total received packets divided by the flow's lifetime. Then we consider the satisfied flows, defined as the ones that complete the file transfer and get at least 95% of the desired minimum throughput (85.5 Kbps). We obtain three performance parameters for an ingress-egress pair: 1) the average "total" satisfied traffic load, which is the aggregated throughput of all satisfied flows, taking into account the minimum and the extra throughput; 2) the average

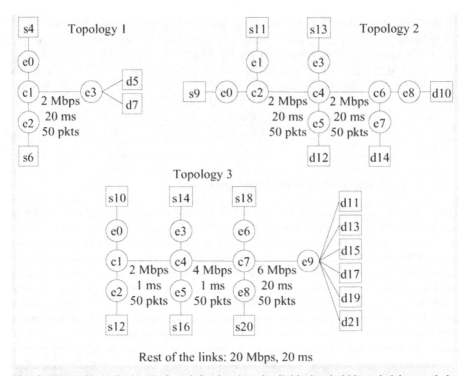

Fig. 2. Network topologies 1, 2 and 3 (showing the link's bandwidth and delay, and the queue's length) used in the simulations

"minimum" satisfied traffic load, which takes into account only the minimum throughput; and 3) the average throughput of satisfied flows (note that it will be always a value above 85.5 Kbps). The first parameter evaluates the total use of resources, the second one its reserved use, and the third one indicates how much extra throughput the satisfied flows get.

The average values of satisfied traffic load are obtained by averaging over the simulation time, but without considering the initial period of the simulation. We make 10 independent replications of each simulation, where independence is accomplished by using different seeds of the random generator (proposed by L'Ecuyer, with code from [15]). The simulation length is chosen so that after the initial transient period a minimum of 10,000 flows generated. From the 10 results we estimate the mean value by computing the sample mean and the 95% confidence interval [15].

3.4 Simulation Results

We compare the results obtained with our scheme with the ideal results that would be obtained by a classical hop-by-hop AC method. We obtain this theoretical result taking into account that in our three topologies the blocking probability in each core link is the same and that the offered traffic for each ingress-egress pair is the same. As a consequence, logical paths with multiple hops obtain a smaller value of satisfied traffic, since flows passing through multiple congested links experience a higher blocking probability. This is a known behavior in classical hop-by-hop AC schemes and end-to-end measurement-based AC schemes [11].

The results obtained with our scheme in topologies 1, 2 and 3 are shown in Figs. 3, 4 and 5, respectively. We show, for an ingress-egress pair, the total satisfied traffic, the minimum satisfied traffic and the average throughput of satisfied flows, versus the average offered traffic ($\lambda\sigma$ bps). The ideal results are represented by dashed lines in the curves of the total satisfied traffic. If the maximum utilization was achieved, the minimum and the total satisfied traffic would be similar and near the ideal result, and the average throughput of satisfied flows would be around 90 Kbps.

In Fig. 3 we compare our scheme (using $T_{03} = T_{23}$ and different measurement durations) with the best-effort service, in topology 1. For clarity we only show the results for the e0-e3 pair, but they are similar for the e2-e3 pair. As expected, in underloading, when resources are enough to satisfy all flows, our scheme achieves the same value for the total satisfied traffic as the best-effort service, which at the same time is equal to the offered traffic (and to the ideal result). In overloading, our scheme achieves higher utilization, which stays almost constant (above 0.9 Mbps, the ideal being 1 Mbps) for the entire range of offered traffic, while the best-effort service achieves utilization tending to zero. On the other hand, the average throughput of satisfied flows decreases for high values of the offered traffic because the unreserved resources decrease.

Another result shown in Fig. 3 is the influence on the performance of the measurement duration $T (= T_{03} = T_{23})$. The best performance is achieved with $T = 0.2$ s. When T is too short (0.04 s) the measurements are wrong and the performance is bad. Moreover, the results show that increasing T (0.5 and 0.9) produces worse performance. The reason for this behavior is that our AC method rejects too many flows when a lot of flows simultaneously meet during the AC phase in a link where there is not enough available throughput for all of them. The available throughput is shared among the flows proportionally to their desired throughput, and the measured throughput of each flow is smaller than the desired one. The AC rejects all flows, and therefore some decisions are erroneous. The number of affected flows depends on how strong the overlapping of the AC phases is, since the more the flows overlap, the more flows are erroneously rejected. An increase in T causes an increase in the number of flows that overlap. This situation also happens when the flow's arrival rate λ increases. Summing up, our scheme obtains better performance with short measurement duration, but there is a limit, because if it is too short, the measurements are wrong.

In Fig. 4 we show the results of our scheme in topology 2. The measurement duration for each ingress-egress pair is $T_{08} = 0.4$ s and $T_{15} = T_{37} = 0.2$ s. The results for the three logical paths are close to the ideal results, although the utilization is a bit smaller.

Finally, in Fig. 5 we show the results of our scheme in topology 3. The measurement duration for each ingress-egress pair is $T_{09} = T_{29} = 0.5$ s, $T_{39} = T_{59} = 0.4$ s, and $T_{69} = T_{89} = 0.2$ s. For clarity we only show the results for the "up" pairs, e0-e9, e3-e9 and e6-e9, but they are similar for the corresponding "down" pairs, e2-e9, e5-e9 and e8-e9. Again, the results for the three logical paths are close to the ideal results, although the utilization is a bit smaller.

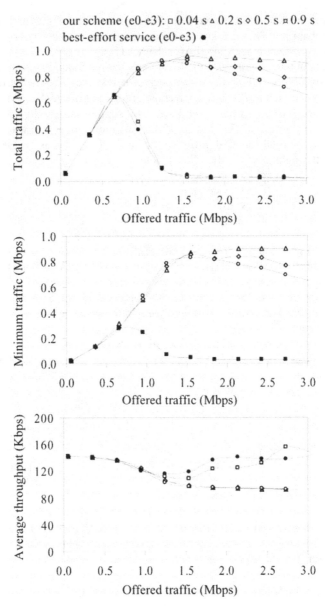

Fig. 3. Our scheme with different T (= T_{03} = T_{23}) and the best-effort service in topology 1

4   Conclusions

We have proposed a new scheme for a network service that guarantees a minimum throughput to flows accepted by AC, suitable for TCP flows. It considers a network path and is able to provide different throughput to different flows as well as protection against non-responsive sources. The scheme is simple and scalable, because it does not need per-flow processing and uses a small set of packet classes in the core, each class having a different discarding priority. The AC method is based on edge-to-edge per-flow throughput measurements using the first packets of the flow.

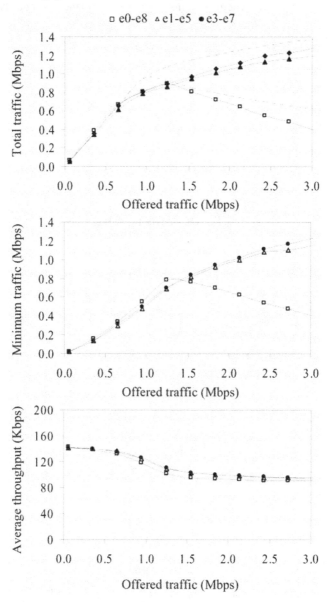

Fig. 4. Our scheme in topology 2 ($T_{08} = 0.4$ s, $T_{15} = T_{37} = 0.2$ s)

We have evaluated the performance through simulations in several topologies with different numbers of hops. We have used different traffic loads of TCP flows carrying files of varying sizes, using a modified source that keeps the short-term sending rate above a minimum value. The results confirm that the scheme guarantees the minimum throughput to the accepted flows and achieves a high value of the utilization of resources, being similar to the ideal results of a classical hop-by-hop AC method. We have also shown that the scheme obtains a better performance using a short measurement time.

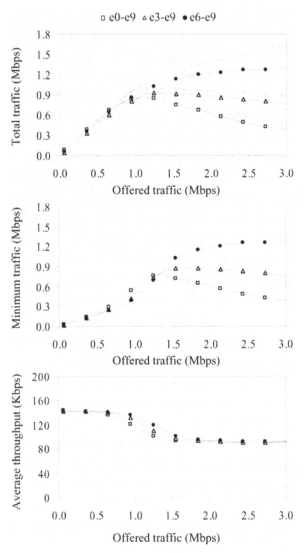

Fig. 5. Our scheme in topology 3 ($T_{09} = T_{29} = 0.5$ s, $T_{39} = T_{59} = 0.4$ s, $T_{69} = T_{89} = 0.2$ s)

References

1. Brownlee, N., Claffy, K. C.: Understanding Internet Traffic Streams: Dragonflies and Tortoises. IEEE Communications Magazine (2002)
2. Jacobson, V.: Congestion Avoidance and Control. ACM Computer Communication Review (1998)
3. Roberts, J. W., Massoulié, L.: Arguments in Favour of Admission Control for TCP flows. 16th ITC (1999)
4. Fàbrega, Ll., Jové, T., Donoso, Y.: Throughput Guarantees for Elastic Flows using End-to-end Admission Control. IFIP/ACM LANC (2003)
5. Clark, D. D., Fang, W.: Explicit Allocation of Best-effort Packet Delivery Service. IEEE/ACM Transactions on Networking (1998)
6. Blake, S., Black, D., Carlson, M., Davies, E., Weiss, W., Wang, Z.: An Architecture for Differentiated Services. RFC 2475 (1998)
7. Mortier, R., Pratt, I., Clark, C., Crosby, S.: Implicit Admission Control. IEEE Journal on Selected Areas in Communications (2000)
8. Kumar, A., Hegde, M., Anand, S.V.R., Bindu, B.N., Thirumurthy, D., Kherani, A.A.: Non-intrusive TCP Connection Admission Control for Bandwidth Management of an Internet Access Link. IEEE Communications Magazine (2000)
9. Fredj, S. B., Oueslati-Boulahia, S., Roberts, J.W.: Measurement-based Admission Control for Elastic Traffic. ITC 17 (2001)
10. Bianchi, G., Capone, A., Petrioli, C.: Throughput Analysis of End-to-end Measurement-based Admission Control in IP. INFOCOM (2000).
11. Breslau, L., Knightly, E., Shenker, S., Stoica, I., Zhang, H.: Endpoint Admission Control: Architectural Issues and Performance. ACM SIGCOMM (2000)
12. Rosen, E., Viswanathan, A., Callon, R.: Multiprotocol Label Switching Architecture. RFC 3031 (2001)
13. UCB/LBL/VINT Network Simulator – ns (version 2)", http://www.isi.edu/nsnam/ns/
14. Fredj, S. B., Bonald, T., Proutière, A., Régnié, G., Roberts, J.W.: Statistical Bandwidth Sharing: a Study of Congestion at Flow Level. ACM SIGCOMM (2001)
15. Law, A. M., Kelton, W. D., "Simulation Modelling and Analysis", MacGraw-Hill, 2000.

Policy-Based Fault Management
for Integrating IP over Optical Networks

Cláudio Carvalho[1], Edmundo Madeira[1],
Fábio Verdi[2], and Maurício Magalhães[2]

[1] Institute of Computing (IC-UNICAMP)
13084-971 Campinas, Brazil
{claudio.carvalho,edmundo}@ic.unicamp.br
[2] DCA-FEEC-UNICAMP, 13083-970 Campinas, Brazil
{verdi,mauricio}@dca.fee.unicamp.br

Abstract. In this paper we present a policy-based architecture for aggregating (grooming) IP/MPLS flows (packet-based LSPs) within lightpaths taking into account the possibility of having to cope with further transport faults. The defined policies try to minimize the negative impact when a failure is detected in the optical transport network. Such policies deal with 1+1, 1:1 and 1:N schemes of protection. In our model, IP/MPLS flows are divided into High Priority (HP) and Low Priority (LP) traffics. The architecture is composed of an Admission Control responsible for receiving the requisitions from the IP/MPLS network and forward them to the Policy Manager which in turn is responsible for applying the policies. The architecture also has a Fault Manager responsible for accounting the failures and a Resource Manager responsible for managing the lightpaths. Our approach has been implemented to validate the policies and the results showed that the defined policies decrease the number of affected LSPs when a given lightpath fails.

1   Introduction

In these last few years, optical networking technology has been considered as a solution for bottlenecks found in today's networks. Typically, these networks have ten to thousands of Gb of available bandwidth and likely consist of elements such as routers, switches, Dense Wavelength Division Multiplexing (DWDM) systems, Add-Drop Multiplexors (ADMs), photonic cross-connects (PXCs) and optical cross-connects (OXCs) [1]. At the same time, due to the advent of Generalized Multiprotocol Label Switching (GMPLS) [1], the provisioning of connections in optical networks can be considered as a partially solved problem.

Although the optical network solves many known problems, it brings new challenges for the research community. One of the main problems deeply analyzed is related to how to minimize the impact of failures in the network. Since each link has a high bandwidth, a failure in a link will cause a lot of data loss. There is much effort in trying to use the same idea of SONET/SDH networks whose time of recovering is about 50 ms. However, it is very difficult to reach such time

T. Magedanz, E.R.M. Madeira, and P. Dini (Eds.): IPOM 2005, LNCS 3751, pp. 88–97, 2005.

in a meshed optical network. The IETF has defined the GMPLS architecture by extending some protocols already used in MPLS networks. These protocols have been defined for dealing with failures treatment. An example of that is the *Notify* message defined in the Resource Reservation Protocol (RSVP) that was extended to support GMPLS networks [2]. There are also some tentatives related to inter-domain protection [3] but nothing is defined as standard yet.

Due to the growing of new optical technologies and its high bandwidth, it is expected that many packet-based network flows will be nested within lightpaths[1] to cross the optical domain and reach their destination. Lightpaths are seen as LSPs (Label Switched Paths) or optical LSPs (from now on optical LSP and lightpath will be used interchangeably) and because of technologies like DWDM it is now possible to have a very large number of parallel links between two adjacent nodes (hundreds of wavelengths, or even thousands of wavelengths if multiple fibers are used).

Although GMPLS considers all the above kinds of data forwarding, the one that is emerging is IP over DWDM networks. In this context, the overlay model is very indicated for service providers (e.g. Telecom companies) since they are the major part interested in acting as transport networks for client IP networks. A very typical and promising scenario is to have MPLS client networks with their packet-based LSPs asking for an optical resource (typically an optical LSP) in order to cross the optical domain and get their destination. Although there is a great interest in the GMPLS architecture, we do no assume that the control plane is based on it. Our approach is general enough and there is no relation to what kind of technology is used in the control plane.

Depending on how the aggregation of packet-based flows within lightpaths is done, the use of the network bandwidth can be maximized or wasted. It is clear that if some rules are followed, the optimization of the network resources is increased and more traffic may be accepted. In this work we are interested in minimizing the impact of failures in the optical domain. The policies we have defined try to aggregate the IP/MPLS traffic in a way that when a given failure happens the number of affected packet-based LSPs is smaller when compared with a scenario without policies. In a previous work [4] we were interested only in maximizing the usage of resources and minimizing the impact of Low Priority LSPs preemptions. In this work, we extended the policies of that work and created new ones to take into account the aggregation of flows within a lightpath to minimize the impact of a failure. The aggregation is dynamically done by the Policy Manager (PM). For each requisition that arrives, the PM looks for a lightpath that can accommodate the flow. If a lightpath is found assuming all the constraints specified by the flow, that flow is then groomed in the lightpath, otherwise the requisition is refused.

The research community has defined (not formally) four main types of protection. The most basic and simplest one is the self-explained unprotected traffic. In the other extreme side is the 1+1 protection. It defines that for each primary

---

[1] The aggregation of lower order LSPs within higher order LSPs is well known as traffic grooming problem

lightpath there is exactly one dedicated backup lightpath carrying the same traffic at the same time. The egress node selects the best signal to be dropped. In case of a failure, only the egress node needs to switchover to the backup. In between these two levels, there are two levels of protection named 1:1 and 1:N. In the 1:1 scheme, the traffic is only sent in the primary lightpath and the backup lightpath can be used for extra traffic. When a failure affects the primary lightpath, the extra traffic being transported on the backup needs to be blocked and the traffic from the primary lightpath is preempted to the backup. The switchover is performed in the ingress node as well as in the egress node. The last scheme of protection is the 1:N. It defines that there is only one backup lightpath for N primary lightpaths. If one of these primary lightpaths comes to fail, the remaining N-1 primary lightpaths become unprotected until the failure is repaired. More details about recovery can be found in [6].

Although some works deal with the grooming and multilayer integration, to the best of our knowledge, none of them addresses the failure problem during the admission of the traffic. In [5], a traffic engineering system is presented considering the multilayer approach and taking into account both methods of routing, off-line and on-line. In [8], the traffic grooming problem is well treated and a formulation on how to use an integer linear programming is presented. This current paper proposes a set of policies to manage the installation and aggregation of packet-based LSPs within optical LSPs assuming that there are several lightpaths between two end nodes and tries to minimize the impact of failures.

This paper is organized as follows. In the next section we describe the architecture and detail the policies that were defined for this work. Section 3 shortly discusses the implementation and the scenario used to validate the policies. Such section is mainly dedicated to show the results obtained in our simulations. Finally, Section 4 concludes the paper and draws some future works.

2 Detailing the Architecture and the Policies

2.1 Architecture

The architecture proposed in this work is composed of five management modules: Admission Control, Fault Manager, Policy Manager, Resource Manager and Policy Repository. These modules were designed in order to get a basic infrastructure to apply policies in optical networks as well as to control all the necessary information for the management of the IP/MPLS over DWDM integration [7]. The architecture is presented in Fig. 1 and in the following we make a brief explanation about each module.

– *Admission Control* (AC): The Admission Control receives the requisitions sent by the IP/MPLS networks and prepare them, loading the lightpaths (from the Resource Manager) between the source/destination pair. After getting all the lightpaths that connect the ingress and the egress nodes, the AC sends such information to the Policy Manager which in turn is responsible for applying the

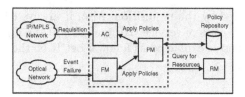

Fig. 1. The proposed Architecture

policies (see below). The AC module is also in charge of re-sending to the Policy Manager the traffic flows that were blocked during the admission phase in a tentative of re-admitting them;

– *Policy Manager* (PM): The Policy Manager implements the policies by analyzing a pool of candidate lightpaths (received from the AC), trying to find one with available resources to accommodate a given IP/MPLS requisition. Also, the PM is responsible for receiving a pool of failed lightpaths from the Fault Manager in order to try to re-admit them by following specific policies to deal with failures;

– *Fault Manager* (FM): The main function of the Fault Manager is to receive the link failure events generated by the optical network equipments and prepare the lightpaths contained in the fiber by separating them in groups of lightpaths according to their type of protection. Then, the FM sends each group of lightpaths to the Policy Manager which in turn applies the specific defined policies for failures treatment;

– *Resource Manager* (RM): The Resource Manager is responsible for storing the information about the virtual and physical topologies. It is accessed by the AC, FM and PM in order for them to obtain any kind of data related to the resources and policies.

2.2   Policies

We developed three groups of policies. Basically, the policies defined in the G1, G2 and G3 groups try to accommodate each IP/MPLS flow within a lightpath. When the failure happens in the transport optical network there is no much effort to be done since the traffic was aggregated during the admission control and now, after the failure, the only procedure that can be done is to preempt the protected flow and, as an extra task, try to re-admit some failed traffic. Note that the tentative of re-admitting traffic is done by re-sending the failed traffic to the PM and let it to apply the policies. In the following, we explain each group of policy separately.

– *Policy Group 1* (G1): This group is the simplest admission policy group. When a requisition arrives in the PM, it tries to install the requisition in a lightpath that offers exactly the same protection as required. It does not consider the class of service of the requisition;

– *Policy Group 2* (G2): It has an intermediate complexity. Its approach is to admit an LSP in a lightpath whose level of protection matches with the level of protection required by the requisition. Also, it always tries to keep together LSPs with the same class of service (HP and LP) in the lightpaths. This group of policies can be better explained as follows: Let R be the Requisition and L a given lightpath.

- if R is Unprotected
  - if R is HP
    1. Aggregate R in an unprotected L if the LSPs already aggregated in L have the same class of service of R;
    2. Aggregate R in an unprotected L that is empty;
    3. Aggregate R in an unprotected L. Probably this L will have both LP and HP LSPs;
    4. Aggregate R in an unprotected L if the removal of one or more LP LSPs of L releases enough bandwidth to install R;
  - if R is LP
    1. Repeat the 3 first steps described above for HP;
    2. Aggregate R in a backup L that is not empty;
    3. Aggregate R in an empty backup L;
    4. Aggregate R in a protected primary L that is not empty. For this condition and the condition five below, L can be an 1:1 or 1:N primary L, but not an 1+1 primary L;
    5. Aggregate R in a protected primary L that is empty;
- if R is 1+1
  1. Aggregate R in an 1+1 primary L that is not empty;
  2. Aggregate R in an 1+1 primary L that is empty;
- if R is 1:1
  1. Aggregate R in an 1:1 primary L that is not empty;
  2. Aggregate R in an 1:1 primary L that is empty;
  3. Aggregate R in an 1:1 primary L if the removal of one or more LP LSPs of L releases enough bandwidth to install R;
- if R is 1:N
  1. Aggregate R in an 1:N primary L that is not empty;
  2. Aggregate R in an 1:N primary L that is empty. For this condition the following rule needs to be accomplished: Let k be equals to the N primaries protected by the backup of L. Then the arithmetic mean of the sharing index among these k lightpaths needs to be lower than the mean of any other different k lightpaths. The sharing index of L indicates the percentage of sharing of its fibers with the other (k-L) lightpaths;
  3. Aggregate R in an 1:N primary L if the removal of one or more LP LSPs of L releases enough bandwidth to install R;

– *Policy Group 3* (G3): Basically, this group of policies performs the same tasks as the G2. However, there are two main differences. The first one is that if the level of protection required by the requisition is not available, this group tries to aggregate the flow in a lightpath with a higher level of protection (if there is one available). This approach is specifically used for 1:N and, as a consequence, the 1:N requisition can be accommodated in an 1:1 lightpath. The second difference is that this group allows to break a given 1:N group to attend 1:1 requisitions. Thus, when an 1:1 requisition arrives and there is no such a level of protection to attend the flow, the policy breaks an 1:N group (if there is one available) in such a way that one of the primary lightpaths of the 1:N group becomes the primary lightpath of the 1:1 level of protection. The backup lightpath of the 1:N becomes the backup of the 1:1 protection. The remaining N-1 primary lightpaths become unprotected. Note that theses two differences are inversely related.

3 Implementation and Results

To test the defined policies, we developed a simulator using the Java language. For sake of space we will not show the policy class diagram. In order to better comprehend the following graphs we firstly show the transition flow that represents the state of an IP/MPLS flow (see Fig. 2).

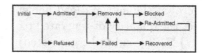

Fig. 2. The Transition Flow of an IP/MPLS Requisition

The initial state represents the arriving of the requisition. From the initial state, the requisition can be admitted or refused. If the requisition is admitted, it can go to the removed state that is an intermediary state whereby a new decision needs to be taken. From that state, the requisition can be blocked (could not be aggregated in another lightpath) or readmitted (the requisition was removed and could be aggregated in another lightpath). From the readmitted state the requisition can be removed again and the loop continues. Back to the admitted state, the requisition can fail (failed state). The failed state means that the requisition is located within a lightpath whose fiber failed. Then it can be recovered which means that it was previously protected and after the failure it was directly switchedover to its backup or, it can be removed (unprotected traffic) continuing the loop as before (from the removed state).

The physical topology used in our simulations is shown in Fig. 3. The lightpaths were created from node 2 to node 6 following different physical routes. Each physical link has two unidirectional fibers (one for each direction) and each fiber has 10 lambdas (wavelengths) with 1 Gb/s in each one. With this

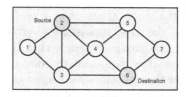

Fig. 3. Physical Topology used in the simulations

physical network, 36 lightpaths (36 Gb/s) from node 2 to node 6 could be created. The quantity of unprotected lightpaths is 4, 1:N is 6, 1:1 is 2 and 1+1 is also 2. For the 1:N scheme of protection we defined 1:3 what means that there are 3 primary lightpaths being protected by 1 backup. This results in 6 groups of 1:3 (6*(1+3)=24). In case of 1:1 and 1+1, for each primary lightpath there is one backup. Thus, since there are two 1:1 and two 1+1 lightpaths, we have the total of 8 lightpaths in these two groups. Then, by summing 24 (1:N) + 8 (1:1 and 1+1) + 4 (unprotected) we have 36 lightpaths.

We have created 8 different traffic loads to validate the policies. From 80% (0.8) to 240% (2.4) of the network bandwidth (36 Gb/s). With these different loads we were able to test the behavior of the policies in scenarios that the quantity of generated traffic is lower than the capacity of the network and to the other extreme, we stressed the network with a high load. The percentage of generated traffic for requisitions (IP/MPLS traffic) for each type of protection is as follows: 35% for unprotected, 15% for 1:N, 20% for 1+1 and 30% for 1:1. Such traffic is generated taking into account the network load percentage. As an example, for 120% (1.2) of traffic load, the quantity of generated requisitions in Gb for 1:1 is: 36 Gb (network capacity) * 1.2 (load to be generated) * 0.3 (percentage of 1:1) $\approx$ 13 Gb/s. The minimum bandwidth required for each requisition is 50 Mb/s and the maximum is 400 Mb/s. Statistically, the average bandwidth for each requisition is then 225 Mb/s. The simulations perform 20 iterations and then the arithmetic mean is obtained. A single fiber failure is randomly generated for each iteration.

Figure 4 shows the quantity of traffic that was admitted in the optical network.

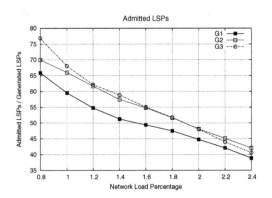

Fig. 4. Percentage of admitted traffic

Note that the G1 is the worst group of policies (actually it is the simplest one). The G3, considered the most sophisticated group, performs better when compared with the other two groups. Observe that G3 and G2 admit basically the same quantity of flows. It is important to point out that the percentage of admission depends on how the requisitions are aggregated within each lightpath. This problem is similar to the knapsack problem [9].

Figure 5 depicts the quantity of admitted traffic specifically for 1:1. While G1 and G2 have about 14.5% of admitted traffic with 80% of traffic load, the G3 has 26%. This difference continues until 240% of traffic load. Remember that the explanation for this good behavior of G3 group is because it breaks the 1:N groups to admit 1:1 traffic (see Section 2.2). Hence, since for our simulations we have generated more 1:1 traffic, the G3 proved to be efficient for this kind of scenario. The G3 group of policies is strongly indicated for scenarios that have 1:N schemes of protection in the optical network and most of the IP/MPLS flows are 1:1.

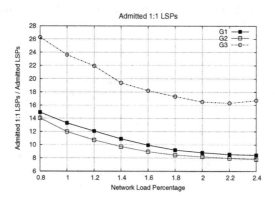

Fig. 5. Percentage of 1:1 admitted traffic

Figure 6 depicts the percentage of failed HP LSPs after the event of a failure. The interpretation of the graph is as follows. After the failure in a fiber, we count how many LSPs (including HPs and LPs) were within that fiber. Then we count how many of them are HPs since the policies always try to save HPs. We can see that G1 performs better than G2 and G3 for all traffic loads, except for those lower than 1.0. Not surprisinlgy, it occurs since the number of HP LSPs admitted in G1 is smaller than the number of HP traffic admitted with G2 and

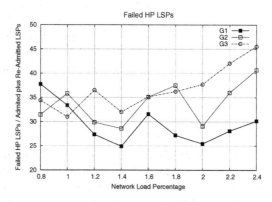

Fig. 6. Percentage of Failed HP LSPs after the failure

G3 (see Fig. 4). A graph, not presented here for sake of space, show that G3 admitted about 50% (HP) of the generated traffic for all traffic loads, and G1, differently, admitted 48% with 0.8 of traffic load and gradually decreases until 37% with 2.0 traffic load.

Figure 7 shows the percentage of LSPs that were blocked after the event of a failure. The G3 group performs better than G2 and G1 from 0.8 to 1.6 traffic loads. Figure 7 should be analysed together with the numbers shown in Fig. 6. Note that as the quantity of failed HPs increases with the traffic load, the quantity of blocked HPs also increases except for G3 from 0.8 to 1.6 of traffic load. This means that the G3 group of policies is able to manage and readmit the HP traffic until 1.6 keeping the quantity of blocked HPs lower than G2 and G1 as desired.

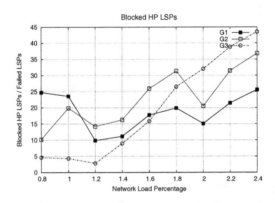

Fig. 7. Percentage of blocked HP LSPs after the event of a failure

The trade off between G2 and G3 can be decided based on specific rules of the optical network provider. As a general rule, if the manager of a given domain is interested in admitting more traffic, mainly 1:1 traffic, than the G3 group should be used. G3 is also indicated if the traffic load is less than 1.6 since in this case the quantity of blocked HPs is lower than G1 and G2 (see Fig. 7). However, if the provider has a traffic load higher than 1.8 and is not interested in prioritising 1:1, than G2 could be used. As a conclusion, if the manager of a given domain has a traffic matrix that forecasts the type and the load of traffic to be admitted in the optical domain, he can better decide on what group of policy to choose.

4   Conclusion

In this paper we presented an architecture for policy-based fault management in optical networks. The policies we defined in this work try to aggregate IP/MPLS flows within lightpaths in way that when a failure happens, the impact of such failure is minimized. The architecture is composed of an Admission Control, a

Policy Manager, a Resource Manager and a Policy Repository. The policies work with the idea that optical networks have a high amount of available bandwidth in each physical link. If such a link comes to fail, the quantity of data that will be lost is consequently very high. Solutions that are only based on schemes of protection such as 1+1, 1:1 and 1:N have been widely discussed. Such solutions can be improved if the type of traffic being transported within a lightpath is considered when aggregating the flows. The policies defined in this paper showed that the number of IP/MPLS flows that are affected when applying the policies is smaller when compared with a scenario that does not use the policies.

As further works we are interested in considering the multi-hop traffic aggregation as well as to explore novel policies for admission control. Also, an important point to be addressed is related to the end-to-end multi-domain connections and Optical VPNs.

Acknowledgments

The authors would like to thank Ericsson Brazil for its support.

References

1. E. Mannie. Generalized Multi-Protocol Label Switching Architecture. RFC 3945, October 2004.
2. L. Berger. Generalized Multi-Protocol Label Switching (GMPLS) Signaling Resource ReserVation Protocol-Traffic Engineering (RSVP-TE) Extensions. RFC 3473, January 2003.
3. J-F. Vasseur and A. Ayyangar. Inter domain GMPLS Traffic Engineering - RSVP-TE extensions. draft-ayyangar-ccamp-inter-domain-rsvp-te-02.txt, January 2005.
4. F. L. Verdi, E. Madeira and M. Magalhães. Policy-based Admission Control in GMPLS Optical Networks. First IEEE Broadnets'04 (formerly OptiComm), San Jose, USA, pages 337–339, October 2004.
5. P. Iovanna, M. Setembre and R. Sabella. A Traffic Engineering System for Multilayer Networks Based on the GMPLS Paradigm. IEEE Network, pages 28–37, March/April 2003.
6. E. Mannie and D. Papadimitriou. Recovery (Protection and Restoration) Terminology for Generalized Multi-Protocol Label Switching (GMPLS). draft-ietf-ccamp-gmpls-recovery-terminology-05.txt, October 2004.
7. F. L. Verdi et al. Web Services-based Provisioning of Connections in GMPLS Optical Networks. The Brazilian Symposium on Computer Networks (SBRC 2005), Fortaleza, Brazil, May 2005.
8. R. Dutta and N. G. Rouskas. Traffic Grooming in WDM Networks: Past and Future. IEEE Network, pages 45-56, November/December 2002.
9. T. H. Cormen. Introduction to Algorithms. Second Edition, The MIT Press.

POBUCS Framework: Integrating Mobility and QoS Management in Next Generation Networks

Fabricio Carvalho de Gouveia and Thomas Magedanz

Technical University of Berlin, Franklinstr. 28-29, D-10587, Berlin, Germany
{gouveia,tm}@cs.tu-berlin.de
http://www.av.tu-berlin.de

Abstract. The Internet is composed by different domains, each of them managed by different ISPs (Internet Service Providers). These Domains have different access networks and capabilities, users and policies that rule their behavior. Policies control QoS and other service parameters like security, and automate network management. Each ISP has an SLA (Service Level Agreement) established between its users, which defines what kind of resources and prices was agreed to be offered to each user. However there is no SLA between a visited domain and a mobile user. Thus there is a need to integrate mobility and QoS management envisaging seamless services and users' satisfaction. We propose POlicy Based Unified access Control System (POBUCS) Framework providing mobility and QoS management, in an integrated way, for IP Multimedia Systems and Next Generation Mobile Networks. The architecture presents the Domain Policy Manager (DPM) as the Decision-making and controller of a domain. The proposed architecture uses policies for inter-domain negotiation.

1 Introduction

The provision of Quality of Service (QoS) is very important for the satisfaction of users and is a major concern in the fields of telecommunication and multimedia networks. With the new emerging area of IP Multimedia Subsystem (IMS) [1] and Next Generation Networks (NGN), QoS and mobility support are essential for the success of such networks.

The concept of NGNs is based on the convergence of fixed and mobile telecommunication networks and the Internet towards an all-IP environment. These NGNs support the provision of integrated information and communication services in face of increasingly more complex value chains, enabled by so-called service delivery platforms (SDPs). Therefore an NGN is an environment of high complexity, in which different actors, such as fixed and mobile network operators, service providers, system integrators and an open set of application providers have to co-operate for the provision of advanced converged services. A converged service is the integration of voice, multimedia, web content and Web Services provided seamlessly over all kinds of access technologies. This heterogeneity increases the complexity and challenges related to the management of such networks. The specific problem addressed by this research is how to ensure necessary levels of service and performance to critical multimedia and data applications, including service guarantees, while automating management. This scenario is illustrated with an inter-domain vertical handover, where

T. Magedanz, E.R.M. Madeira, and P. Dini (Eds.): IPOM 2005, LNCS 3751, pp. 98–107, 2005.

the QoS must be guaranteed with an acceptable level while a mobile user roams through domains.

To cope with these requirements, we propose the POlicy Based Unified access Control System (POBUCS) Framework. POBUCS is a management framework that addresses end-to-end QoS in the case of user mobility. This integration is managed by policies based on resources negotiation among domains.

The content of the paper is structured as follows: First we discuss briefly IMS playground within the FOKUS 3Gb Testbed [2], a platform for testing and validating the underlying framework. In section 3 we analyse the requirements that NGNs impose on the management plane. Related work is commented in section 4. In section 5 we present the POBUCS framework model, components and its capabilities. An application scenario is described in section 6, and finally, the conclusions and future work are presented in section 7.

2   IMS Playground at Fraunhofer FOKUS

FOKUS 3Gb Testbed consists on three logical layers as shown in Figure 1. The lowest layer is the network plane that integrates wireless and fixed access networks, such as GSM, WLAN, digital video broadcasting, UMTS, etc. including the related end systems. Next to the network layer is the control and management layer, constituted by signaling and management components. This layer is responsible to provide the IMS capabilities like QoS, security, AAA, etc. forming the service delivery platform. The highest layer is the application layer, where the services are developed.

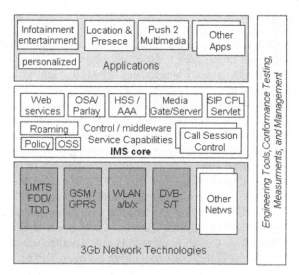

Fig. 1. Open IMS within the 3Gb FOKUS Testbed

FOKUS researches and develops the building blocks needed both for end-to-end seamless integration of technologies and end devices for the deployment of open flexible communication services and applications. The core elements of the 3G beyond Testbed comprise of play grounds for Multimedia Value-added Service Platforms covering the whole range of relevant telecommunication service platforms for converging networks, like OSA/Parlay and IP Multimedia Systems. The Open IMS provides a platform for the validation of the POBUCS framework.

3   Requirements for the Management of Next Generation Networks

As discussed in the introduction, NGNs puts new burdens in the management of these kinds of heterogeneous networks (different wired and wireless technologies). This section summarizes the requirements for the management of QoS and Mobility:

QoS Requirements

- To guarantee QoS among different domains and networks;
- End user perception of QoS.

Mobility Requirements

- ability to change access point and/or terminal;
- ability to get access from any network access point, including all access technologies identified;
- ability to get services in a consistent manner, subject to the constraints experienced in their current situations;
- The Framework must be aware of user availability.

4   Related Work

In Recent years Policy-based management has become a great topic of research. However, most research in this area focus in intra-domain QoS or Security. There is no much work with Policy Based Management (PBM) that addresses mobility issues. A Study in 3GPP [3] address the WLAN-3GPP system interworking and specifies 6 scenarios with different levels of integration. Policies are used to control the bearer traffic on 3GPP IMS. PBM addressing mobility hasn't been the subject of as many works as the security or QoS ones. It is difficult to meet all the requirements for automated management in a heterogeneous system. However, there are some works trying to define a model to use policies in the mobility management. In [4] the authors propose extensions to the COPS protocol, called COPS-MU (Mobile User) and COPS-MT (Mobile Terminal), to deal with terminal and user mobility. They define new policy objects in the COPS protocol for terminal and user registration. There is another extension [5] called COPS-SLS (Service Level Specification) to support the QoS negotiation. We claim that an integrated Policy Architecture can support mobility and QoS management if there is cooperation between DPMs. The DPM can be compared to the Bandwidth Broker (BB) in the Differentiated Services (Diffserv) Architecture [6]. The European research projects AQUILA [11] and TEQUILA [12] use a centralized QoS approach like the DPM. With a single control point, the integration with different mobility schemes and location management becomes more flexible [13] The DPM is a centralized manager that handles the resources of one domain. It keeps the actual stand of resources and reservations in its domain. They motivate to use a central approach in order to support anticipated handover with pre-reservations, allowing the mobile node to perform handover only if enough resources are available.

Another work [20] proposes PBM to manage QoS, AAA and mobility, extending the policy-based networking to the user terminal. However this approach can introduce performance problems depends on whether the end-user network is wired or wireless environment. In a wireless case the PEP in the terminal interacts with the PDP introducing overload information, what is very important to consider since the radio resources are scarcer than in a wired environment. All PBM works evaluated, manage the QoS at the IP level. We have also chosen this approach, and to reach coordination in the integration of both we propose to manage mobility at the same level. For such mechanisms to be efficient, we need an efficient mobility management scheme optimized for QoS. Integration of mobility management with QoS can bring advantages to the system, for instance, it is possible to take into account a richer set of parameters to initiate hand-offs. The decision to switch to another cell can be made not only based on the signal to noise ratio, but also on the current load in a cell, the level of available resources and the state of pre-reservations or administrative policies. This integration envisages the improvement of the overall end-to-end performance and the reduction of management entities in the network.

5 POBUCS Framework

The POBUCS framework is based on policies. This section describes the features of POBUCS and its approaches.

Policy-based management is the application of the concepts of organizational policy, like agreements and procedures, to the governance of computer-based systems. The problem with current network management systems is that they lack the ability to state long-term, network-wide configuration objectives, and have them automatically realized in the network. A policy-based management system allows the network operator to enter the above objectives as policies into the management system, and ensures automatic enforcement of these policies so that no further manual action is required on the part of the network operator.

In a standard policy-based network, policies consist of a set of conditions and a set of actions (eg. If <condition> then <action>) [7]. Table 1 shows the categories of policies.

Table 1. Categories of policies

Policy Categories	Description
Authorization [8]	Permit or forbid an autonomous entity from carrying out an action
Obligation [8]	Defines the duties, roles and responsibilities of an autonomous entity
Abstract [9]	Specifies a goal, objective or constraint that needs to be achieved without specifying how it is to be achieved
Concrete [9]	Specifies a process or procedure that explicitly needs to be followed

The advantages of concrete policies include that they are typically easy to achieve, because there are no details which are unclear or unspecified. Furthermore, the manner is which the outcome is measured is also specified, so that autonomous entity knows that it has met its obligation. Abstract policies have missing details, requiring enough knowledge and intelligence from the autonomous entity to work out how it can be achieved, and how the outcome can be measured. On the other hand, the abstract policies are easier to read and write and can be applied to more general (or aggregated) properties of the system [10]. Examples of Typical usage of policies include device configuration, Service Level Agreement (SLA) between two adjacent domains, resource reservation, admission control and charge correlation.

The POBUCS uses policies focusing QoS and mobility integration. This allows an operator to specify how the network is to be configured and monitored through rules that will trigger when determined events occur. For example, an abstracted policy [14]:

Jack has access to the high quality Video on Demand service

The high quality Video on Demand service must have higher priority than best-effort traffic could be represented by the following rules in a hypothetical policy language (which contains both 'data' and 'behavioural' statements):

Jack.Services = [HighQualVoD]
HighQualVoD.priority > BestEffort.priority

This is a simple example. A more refined rule could be something like that (for diffserv case) [14]:

Jack.IPAddress = 192.168.0.1
HighQualVoD.Port = 1024
HighQualVoD.DiffServCodePoint = AssuredForwarding11
AssuredForwarding11.DSCP = 001010b
AssuredForwarding11.Scheduling = PriorityQueuing
EdgeRouters = [192.168.1.1, 192.168.5.1]
CoreRouters = [192.168.2.1, 192.168.3.1, 192.168.4.1]

And the resulting concrete policy:

For all [EdgeRouters]
If UDPPacket.DestIPAddress = 192.168.0.1 and UDPPacket.Port=1024
Then UDPPacket.DSCP = AssuredForwarding11.DSCP
For all [CoreRouters]
If IPPacket.DSCP = AssuredForwarding11.DSCP
Then Queue1.Enqueue (IPPacket)

The use of policies brings many advantages:

- It allows the automation of many management tasks, that can be too difficult or numerous.
- It introduces the capability of the management system to adapt its behavior dynamically to different environmental conditions [14].
- It allows the system to achieve higher-level objectives, such as business goals/rules and service-level management [15].
- The establishment of configuration management capabilities into the network management space. In addition, the policy system can remain online during policy changes [16], improving network availability and adaptability.

5.1 Conceptual Model of POBUCS

The reference model of POBUCS framework consists of two main elements, the Domain Policy Manager (DPM) acting as a Policy Decision Point (PDP) and the Policy Enforcement Point (PEP), and is depicted in Figure 2. The components of this conceptual model structure are based in the Internet Engineering Task Force (IETF) policy framework group [9].

The main task of the DPM is to enable coordination between events in the application layer and the resource management in the IP bearer layer during session establishment. The DPM weighs the policy request sent by the PEP, as a result of a policy event against a corresponding set of policy rules. As a response to a policy request, the PDP either evaluates the policy rules for the request, which is referred to outsourced policy or retrieves the set of policy rules relevant for the request, which is referred to provisioned policy (The Provisioning and Outsourcing policy models are described in [17]). The policy decision or the set of policy rules is then transported to a PEP using the policy transaction protocol Common Open Policy Service (COPS) protocol. In POBUCS, the DPM is the final authority all PEPs need to refer for actions to be taken (Decision-making).

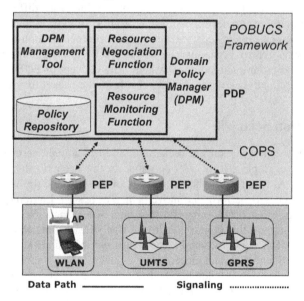

Fig. 2. Conceptual Model of POBUCS

The role of the PEP is to ensure that only authorized IP flows are allowed to use network resources that have been reserved and allocated to them. The PEP in the Access Router (AR) is responsible to drop the IP flow that was not permitted by the DPM. This process is called policy-based admission control. This process ensures that an IP flow is only allowed to use resources that have been approved by the policy rules. The PEP may store decisions in a local PDP. In this case, the AR can make admission control decisions without additional interactions with the DPM, reducing this way, the traffic between them and lessening the processing load on the DPM.

The modules of the DPM are:

- **DPM Management Tool:** Is the interface for the domain administrator to add, edit and remove policies, as well as to control the Policy Repository for consistency and search for policies conflicts.
- **Policy Repository:** Is the entity where all the policies of a domain are stored (for home users, as well as policies that rule what to do when a foreign user joins the domain).
- **Resource Negotiation Function:** Is responsible to reach an agreement for requests that the domain can't provide for a roaming user. It searches for common parameters between the domain and the user.
- **Resource Monitoring Function:** Monitors the network for the current status of resources and ensures that the contracted services are being met.

The interoperability of POBUCS with other networks is also considered. In an end-to end scenario there are several administrative domains, each with its own policies, resource management architectures and traffic mechanisms. A common protocol would be needed for communicating end to end the QoS requirements of user traffic, while at the same time respecting the individualities of the autonomous operation of each traversed domain [18]. However it is not possible with different autonomous networks and many proprietary solutions. To find a solution to interoperation of the system, it is indispensable to separate the signaling protocol from the carried information. In this way, control signaling can he carried by any signaling protocol, while being understood by any autonomous system. The concept of POBUCS follows this standard, decoupling the network control plane from the packet forwarding plane.

6 Application Scenario

In this section, we give an example scenario of an inter-domain vertical handover (see Figure 3). This example covers IMS domains. In this scenario there are 2 IMS domains: Domain A, which is the home network of the user and the visited Domain B, in which the user will perform the movement. The User's Equipment UE is a Personal Digital Assistant (PDA) attached to a dual access card (WLAN - GPRS) and therefore it can support this type of handover. The SIP signaling IMS and registration process in the IMS is not considered here, but is described in [19]. The focus is on the inter-domain QoS signaling and mobility.

As the user is in his home domain, he has a Service Level Agreement (SLA) with his provider and requests the contracted resources, for instance, he wants to make a VoIP call. The DPM process the call admission control, reserves the required resources and authorizes the session in the local domain. Then, he starts the movement to the Domain B. The DPM in the home domain sends a service request message to the other DPM with the format of the QoS request, such as the type of service (VoIP) and QoS requirements that were authorized in the home services. The DPM on Domain B receives this request and access the Resource Monitoring Function, in order to know if there are sufficient resources to reserve for the user. If not, it informs the previous DPM about this state. If the resources are enough, the Visited DPM will check its Policy Repository to check the rules that will apply in this domain. If there are parameters that must be changed in Domain 2 (The technologies are different),

the Resource Negotiation Function will start the negotiation with the DPM of Domain A. If an agreement is met, the DPM of Domain B will make the local decision
based on the incoming service request message. The PEP in Domain B will enforce
these policies and the handover will be performed. Immediately the PDP of Domain
B will send a Report State message back to Domain A. As the DPM receives this
message, the whole procedure finishes. Figure 4 shows the message flows for the
inter-domain negotiation of the scenario (It doesn't show the authentication of the
user in the visited domain).

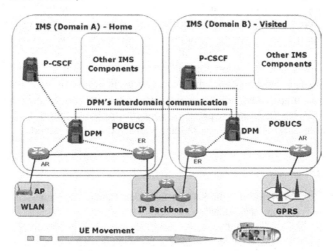

Fig. 3. Inter-domain vertical Handover

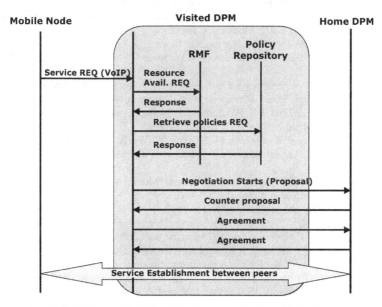

Fig. 4. Message Flows of the Inter-domain Communication

7 Concluding Remarks and Future Work

Along this paper we presented the POBUCS framework for the management of Next Generation Networks and IP Multimedia Systems. This integrated framework addresses mainly three requirements of NGNs: end-to-end QoS, automation of the system and mobility management at the IP level. The management of NGNs needs abstraction to handle heterogeneity, and we claim that policy-based management can bring the sufficient level of abstraction and automation needed in this field. The framework approach uses a centralized manager (one Domain Policy Manager per domain), which makes this framework easier to integrate with different mobility schemes in the literature. Another feature is the decoupling of the control plane from the packet forwarding plane. This allows interoperability with other autonomous systems because the information control can be carried by any signaling protocol. We demonstrated the functionality of the proposed architecture in an inter-domain vertical handover scenario. The POBUCS framework is part of the R&D 3Gb Testbed at Fraunhofer FOKUS Institute. A proof-of-concept implementation is being developed and programmed and will be integrated and validated in the Testbed.

References

1. 3GPP, TS 23.228. IP Multimedia Subsystem; (stage 2), may 2005. www.3gpp.org
2. www.fokus.fraunhofer.de/national_host
3. 3GPP, TS 22.934, Requirements on 3GPP system to Wireless Local Area Network (WLAN) interworking, September 2004. www.3gpp.org
4. G. Pujolle, and H. Chaouchi. "A New Policy Based Management of Mobile Users", Networking 2002, LNCS 2345, pp. 1099-1104, 2002.
5. G. Pujolle, and H. Chaouchi. "QoS, Security, and Mobility Management for Fixed and Wireless Networks under Policy-Based Techniques". IFIP World Computer Congress, 17th edition, August 25-30, 2002, Montréal, Canada.
6. S. Giordano, S. Salsano, S.V. Bergue, G. Ventre, D. Giannakopoulos, "Advanced QoS Provisioning in IP Networks: the European Premium IP Projects." IEEE Com. Mag., January 2003.
7. IETF RFC 3198, "Terminology for Policy-Based Management", November 2001.
8. M. Sloman, "Policy Based Management of Telecommunication Systems and Networks 1998," presented at First UK Programmable Networks and Telecommunications Workshop, Hewlett-Packard Laboratories, Bristol, 1998.
9. draft-ietf-policy-framework-00.txt, M. Stevens, W. Weiss, H. Mahon, B. Moore, J. Strassner, G. Waters, A. Westerinen, and J. Wheeler, "Policy Framework," Policy Framework, 13 Sept 99.
10. N. Sheridan-Smith, D. Colquitt, and J. Wootton, "Moving from Next -Generation Networks to Enriched-Experience Networks," University of Technology, Sydney, Australia, Confidential deliverable 1A, 30 September 2003.
11. T. Engel et al., "AQUILA: Adaptive Resource Control for QoS Using an IP-Based Layered Architecture," IEEE Communications Magazine, Jan. 2003.
12. E. Mykoniati et al., "Admission Control for Providing QoS in DiffServ IP Networks: the TEQUILA Approach," IEEE Communications Magazine, Jan. 2003.
13. N.S. Smith, "A Distributed Policy-based Network Management (PBNM) system for Enriched Experience Networks™ (EENs)", Doctoral Assessment, November 2003.
14. M. Sloman and E. Lupu, "Security and Management Policy Specification," IEEE Network, pp. 10-19, 2002.

15. M.L. Stevens and W.J. Weiss, "Policy-based Management for IP Networks," Bell Labs Technical Journal, pp. 75-94, 1999.
16. M. Sloman, "Policy Based Management of Telecommunication Systems and Networks 1998," presented at First UK Programmable Networks and Telecommunications Workshop, Hewlett-Packard Laboratories, Bristol, 1998.
17. ETF RFC 2748, The COPS (Common Open Policy Service) Protocol", January 2000.
18. S.I. Maniatis, E. G. Nikolouzou, L. S. Venieris, "End-to-End QoS Specification Issues in the Converged All-IP Wired and Wireless Environment", IEEE Communications Magazine, June 2004..
19. 3GPP TS 24.228, "Signalling flows for the IP multimedia call control based on Session Initiation Protocol (SIP) and Session Description Protocol (SDP)"; Stage 3, April 2005.
20. H. Chaouchi. "A New Policy-aware Terminal for QoS, AAA and Mobility Management", International Journal of Network Management 2004, 14: 77-87, John Wiley & Sons, Ltd.

Executable Graphics for PBNM

Rui Lopes[1], Nuno Raimundo[1], Maria Varanda[1],
José Oliveira[2], and Vitor Roque[3]

[1] Polytechnic Institute of Bragança, 5301-854 Bragança, Portugal
{rlopes,neves,mjoao}@ipb.pt
[2] University of Aveiro 3810-193 Aveiro, Portugal
jlo@det.ua.pt
[3] Polytechnic Institute of Guarda 6301-559 Guarda, Portugal
vitor.roque@ipg.pt

Abstract. The specification of a policy is performed in a policy language, usually following a textual representation. However, humans process images faster than text and they are prepared to process information presented in two or more dimensions: sometimes it is easier to explain things using figures and their graphical relations than writing textual representations.

This paper describes a visual language, in the form of graphics that are executed in a networking environment, to define a network management policy. This approach allows to map visual tokens and corresponding arrangements into other languages to which a mapping is defined.

1 Introduction

Network management has been a constant worry among organisations and network operators in the last decades. We have seen several approaches being developed and proposed, from distributed systems solutions (such as CORBA – JIDM, for example – or Mobile Agent based solutions) to specific solutions, such as the SNMP or CMIP. Among them, the SNMP model has, probably, been the most well known and widely used. However, none of them has fully satisfied the community which is still searching for an appropriate model or paradigm that can be used efficiently in network management scenarios.

Policy-Based Network Management (PBNM) has become a promising solution for managing enterprise-wide networks and distributed systems. It is targeted to systems that are dynamic in nature and where stopping and recoding is undesirable – changing the policy rules allows the system to modify its behaviour [1].

The PBNM paradigm has proved, at least in theory, that is a good solution for network management. The ideas involved in this paradigm helps greatly network managers in the complicated tasks of the network administration. In fact, when all the theoretical PBNM concepts are applied to practical and effective management applications, the network administration task will be much easier in terms of time, money and difficulty.

Central to the PBNM is the concept of *policy*, usually considered the link between high-level business specification of desired services and low-level device

T. Magedanz, E.R.M. Madeira, and P. Dini (Eds.): IPOM 2005, LNCS 3751, pp. 108–117, 2005.

configurations that provide those services [2]. This definition implies some form of communication and, consequently, a means of describing the concepts associated with the business-level goals – a *policy description language*.

Currently, there are several languages that can be used in policy definition. Some authors have already discussed some of them and have also presented new approaches [3,4]. However, most of the approaches rely on formalisms that are hard to remember and, sometimes, hard to use, forcing the user to learn new terms and even new constructions. Much of this effort can be reduced by using visual languages, where the user combines pictorial elements to build a flowchart like arrangement or other similar structure [5].

In this paper, we propose the construction of a graphical editor to describe policies. The idea is to create a visual language to specify policies and than use an editor for generating the data structure that can be recognised by the network as a generic representation of a policy. In this context, we are using the PCIM formalism to store and to install the policy in the network.

2   Visual Languages for PBNM

The purpose of a policy description language is to translate from a business specification, such as those found in a Service Level Agreement (SLA), to a common vendor and device-independent intermediate form.

Although there are several approaches and formalisms for specifying policies, there is a common understanding on the concepts involved [1,6,3]. This implicit "understanding" allows the specification of a common representation of policies. In general, we can classify policies in two broad classes: *configuration* policies and *management* policies. Configuration policies are used to define initial, or otherwise *condition independent*, rules, used in the configuration of resources and in the definition of state independent policies. As an example, consider the following sentence:

$$\text{``file 'X' can be accessed by users from group 'students' ''} \tag{1}$$

Management policies can be used to define adaptable management actions, usually based on *event-triggered, condition-action* rules. For example:

$$\text{``notify the admin when the error rate of outgoing packets is increasing''} \tag{2}$$

More focused in the latter approach, the IETF together with the DMTF, has done a remarkable work with the Policy Core Information Model (PCIM) [7] and corresponding extensions – PCIMe [8]. The Policy Framework WG defines *policy* as an aggregation of policy rules [2]. Each policy rule is made up of a set of conditions and a corresponding set of actions. Although this type of policy rule does not explicitly specify an event to trigger the execution of the actions, it assumes an implicit event, such as a process being lauched or a particular traffic flow. In this case, the rules will include an event part [9]:

$$[(policyEvent) \text{ causes}] \ (policyAction) \text{ if } (policyCondition) \tag{3}$$

Grouping is also implicit, as it represents the aggregation of related objects. Finally, each network component may represent or act in a role or in a set of roles. The role is a label which indicates a function of an interface or device. In other words, a policy is a set of rules, which can belong to one ou more groups, and which can be applied to devices acting under a specific role.

2.1   Using PBNM Languages

At the highest level, policies may resemble a human language, such as: "user xpto may transfer files". At system level the same policy may assume a different form, still device and technology independent. Usually, policies are specified in a format which is relatively easy to convert to network configuration commands. This paper does not intend to be a survey of the existing approaches, however it is important to provide some insight into how some of the languages are and how do they look like.

Several languages have been developed, resulting from the effort of the academia as well as private enterprises, such as IBM or Sun. Probably, one of the most popular is the Ponder Policy Specification Language [10]. Ponder is a declarative, object-oriented language that can be used to specify both configuration and management policies. It supports obligation policies that are event triggered condition-action rules for policy based management of networks and distributed systems. Key concepts of the language include domains, to group the object to which policies apply, roles to group policies relating to a position in an organisation, relationships to define interactions between roles and management structures to define a configuration of roles and relationships pertaining to an organisational unit such as a department.

The following example shows how Ponder can be used in policy description:

```
inst auth+ filter {
  subject /Agroup + /Bgroup;
  target UAStaff - DETUAgroup;
  action VideoConf(BW, Priority)
    {in BW=2; in Priority=3;} // default filter
    if(time.after("1900")) {in BW=3; in Priority=1;}
}
```

The above policy says that the members of the predefined groups `Agroup` and `Bgroup` may use the video-conference service to the group `UAStaff` excluding `DETUAgroup`. If the service is used after 7 PM, the bandwidth is set to 3Mb/s with priority 1. In other circumstances, the bandwidth is 2Mb/s and the priority is 3.

XACML [11] is an XML specification for expressing policies mainly dedicated to access control and is being defined by the Organisation for the Advancement of Structured Information Standards (OASIS). The language supports roles, which are the same as groups, and are defined as collections of attributes relevant to a principal. It includes conditional authorization policies, as well as policies with external post-conditions to specify actions that must be executed prior to permitting an access.

The following example describes a scenario where a set of video streaming servers offers tutorials to registered and unregistered users. Registered users have permission to access any server offering a service without time restrictions. Unregistered users can have access to the video-streaming service only from the internal network and not in business-time [12].

```
<service serviceId="TutorialVideoStreaming">
 <description>tutorial video-stream</description>
 <sap>
  <inetaddress> 192.168.200.10 </inetaddress>
  <inetaddress> 192.168.5.3 </inetaddress>
  <protocol>tcp</protocol>
  <port>8976</port>
 </sap>
 <serviceLevel serviceId="Gold">
  <ResourceRsvp AttributeId="qosG7" RsvpClass="G7">
   <TspecBucketRate_r>9250</TspecBucketRate_r>
   <TspecBucketSize_b>680</TspecBucketSize_b>
   <TspecPeakRate_p>13875</TspecPeakRate_p>
   <TspecMinPoliceUnit_m>340</TspecMinPoliceUnit_m>
   <TspecMaxPacketSize_M>340</TspecMaxPacketSize_M>
   <RsvpService>Guaranteed</RsvpService>
   <RsvpStyle>FF</RsvpStyle>
  </ResourceRsvp>
 </serviceLevel>
 <serviceLevel serviceId="Silver"> . . . </serviceLevel>
 <serviceLevel serviceId="Bronze"> . . . </serviceLevel>
</service>
```

The Routing Policy System WG of the IETF has defined the RPSL (Routing Policy Specification Language) [13]. It was one of the first languages for specifying routing policies and aims at generating router configuration from the policy specification [14]. A possible example is:

```
aut-num:      AS2
as-name:      CAT-NET
descry:       Teste
import:       from AS1 accept ANY
import:       from AS3 accept <^AS3+S>
export:       to AS3 announce ANY
export:       to AS1 announce AS2 AS3
admin-c:      AO36-RIPE
tech-c:       CO19-RIPE
mnt-by:       OPS4-RIPE
changed:      estig@ipb.pt
source:       RIPE
```

The IETF did not define a specific language to express network policies but rather a generic object-oriented information model for representing policy information. An advantage of the information modelling approach followed by the

IETF is that the model can be easily mapped to structured specifications such as XML, which can then be used for policy analysis as well as distribution of policies across networks. The mapping of CIM to XML is already undertaken within the DMTF [15]. The IETF has defined a mapping of the PCIM to a form that can be implemented in a directory that uses LDAP as its access protocol [16]. Considering the following example, we will try to specify the attributes of all the necessary PCIM classes.

```
Group students: Role=[studentPrinters] {
  if (studentPrinterQuota < 0) {
    deny printing job;
  }
}
```

The first thing to do is to create an instance of CIM_PolicyRule (Table 1) to define the base of the policy.

Then, the condition is created by defining an instance of CIM_VendorPolicy Condition (Table 2).

After creating the condition, it must be associated to the policy with an instance of CIM_PolicyConditionInPolicyRule (Table 3).

Now the action is specified by creating an instance of CIM_ActionPolicy Condition (Table 4) and it is associated to the policy through an instance of CIM_PolicyActionInPolicyRule (Table 5). The resulting policy must be associated to a group, so we need to create the group (Table 6) and create the association object (Table 7). Finally, the policy must be set to a role (Tables 8 and 9).

Table 1. PolicyRule

CIM_PolicyRule	
Caption	"Policy"
CommonName	"printer quota"
ConditionListType	'DNF'
CreationClassName	"CIM_PolicyRule"
Description	"Controls printing jobs"
ElementName	"printer quota"
Enabled	'Enabled'
ExecutionStrategy	'Do Until Failure'
PolicyDecisionStrategy	'First Matching'
PolicyRuleName	"printer quota"
RuleUsage	"Test the printer quota"
SequencedActions	'Don't Care'
PolicyKeywords	'USAGE'

Table 2. VendorPolicyCondition

CIM_VendorPolicyCondition	
Caption	"Condition"
CommonName	"printer quota"
ConstraintEnconding	"UTF-8"
CreationClassName	"VendorPolicyCondition"
Description	"Test students print jobs"
ElementName	"Test print quota"
PolicyConditionName	"Test print quota"
PolRuleCrtnClassName	"CIM_PolicyRule"
PolicyRuleName	"printer quota"
Constraint	"studentPrinterQuota<0"
PolicyKeywords	'USAGE'

Table 3. PolicyConditionInPolicyRule

CIM_PolicyConditionInPolicyRule	
ConditionNegated	'FALSE'
GroupNumber	1
GroupComponent	"printer quota"
PartComponent	"Test print quota"

Table 4. VendorPolicyAction

CIM_VendorPolicyAction	
ActionEncoding	"UTF-8"
Caption	"Action"
CommonName	"printer quota"
CreationClassName	"CIM_VendorPolicyAction"
Description	"Deny printing job"
DoActionLogging	"Denying printing job"
ElementName	"Deny printing"
PolicyActionName	"Deny printing"
PolRuleCrtnClassName	"CIM_PolicyRule"
PolicyRuleName	"printer quota"
ActionData	"deny printing job"
PolicyKeywords	'USAGE'

Table 5. PolicyActionInPolicyRule

CIM_PolicyActionInPolicyRule	
ActionOrder	1
GroupComponent	"printer quota"
PartComponent	"Deny printing"

Table 6. PolicyGroup

CIM_PolicyGroup	
Caption	"Policy Group"
CommonName	"printer quota"
CreationClassName	"CIM_PolicyGroup"
Description	"policy group"
ElementName	"Students"
Enabled	'Enabled'
PolicyDecisionStrategy	'First Matching'
PolicyGroupName	"Students"
PolicyKeywords	'USAGE'

Table 7. PolicySetComponent

CIM_PolicySetComponent	
Priority	4
GroupComponent	"printer quota"
PartComponent	"Students"

Table 8. PolicyRoleCollection

CIM_PolicyRoleCollection	
Caption	Role
Description	"printers"
ElementName	"Role1"
InstanceID	"IPB:role1"
PolicyRole	"studentPrinters"

Table 9. PolicySetInRoleCollection

CIM_PolicySetInRoleCollection	
Collection	"IPB:role1"
Member	"printer quota"

A clear message resulting from the above paragraphs is that the textual description of policies is a boring and error prone task. With these semantics, the primary goal of PBNM may be compromised since the idea is to simplify tasks through the use of high level, business oriented, declarations. The assistance of a graphical user interface, that can help in filling the various attributes, may simplify the job and may allow a faster definition of policies. We can establish a parallel between the Visual Basic programming environment, where the development of applications is faster and may be achieved by less skilled programmers, and a C++ programming environment, where the development of applications typically takes more time but results in a better structured code.

2.2   Using Visual Languages

A visual language is characterized by the use of graphical notation – a graphical vocabulary and sentences constructed by a spacial combination of symbols (two dimensions). There are several reasons to use visual languages. The basic idea is that humans process images faster than text and that they are prepared to process information presented in two or more dimensions.

The human reasoning is image-oriented: sometimes it is easier to explain things using figures and their graphical relations than writing textual representations. A text has a sequential structure and its visual aspect is always the same. When we read a text we must understand each character to read a word and we must understand all the words to read and understand sentences.

In visual languages each figure can give different information depending on its size, on its colour, on its form and so on. A visual sentence can be more attractive and easier to understand. So, it is common to use drawings to explain things. Many pedagogical tools use visualisations and animations as a simple way to achieve their purposes. However, visual languages are not as simple to specify, process or traduce by automatic mechanisms than textual languages.

There are some visual language compiler generators and other interesting tools (some of them are very specific) that can be used to process visual languages in order to generate other representations.

The idea of using a visual language to specify policies allows the user to work with a graphical editor to build a structure that can be interpreted as a policy. Just like with textual languages, where compiler-compilers, such as LEX and Yacc, build the code to deal with some programmer defined language, graphical languages can also be achieved by compiling a grammar to build an editor automatically.

3   Executable Graphics for PBNM

In the previous sections we described some approaches to specify policies through the use of textual languages. This section presents a visual language, composed of graphical elements which are executed into a textual description. Because we like to interpret the visual richness of objects as executable charts we use the term *executable graphics* [17].

The main idea is to describe the policy language using a grammar, to be compiled into a graphical editor, and the appearance of the tokens using images: The advantages of formal specification of programming language semantics are well known: first, the meaning of a program is precisely and unambiguously defined; second, it offers a unique possibility for automatic generation of compilers or interpreters.

```
screen     : policy resources
resources  : resources resource
           | resource
resource   : NAME roles
policy     : rule roles
roles      : roles role
           | role
role       : STRING
rule       : IF condition THEN actions
           | IF condition THEN actions WHEN events
condition  : dnf | cnf
dnf        : exp1
           | dnf OR exp1
cnf        : exp2
```

```
           | cnf AND exp2
exp1       : cterm AND cterm
exp2       : cterm OR cterm
cterm      : STRING
           | NOT STRING
actions    : actions action
           | action
action     : STRING
events     : events event
           | event
event      : STRING
```

We have chosen a spatial paradigm for the policy definition instead of a chart like approach because we think it is easier to fill spaces than connecting lines. The user will need to drag and drop elements to specific spaces in the user interface, thus making the definition of policies easier. We assume that the non-terminal symbols (`role`, `cterm`, `action` and `event`) are represented as blocks in the user interface in a corresponding pallet. This tablet is the initial drag point and the source for the conditions, actions, events and roles.

A condition is a combination of condition terms (`cterm`) in CNF or DNF form. Considering that we want to build a conjunction of two conditions (*Condition*1 ∧ *Condition*2) it is necessary to drag and drop both of them into the same screen space. The user interface will show the resulting condition inside the same block. If the condition is negated, it is shown as a patterned box with diagonal lines. If we further want to have the result of a conjunction of three more conditions in DNF form, we need to build a disjunction of the resulting blocks (Figure 1). The colours represent the type of operation. A light colour (light gray, in this case) represents the AND operation and the darker colour represents the OR operation. Using CNF instead of DNF would reverse the colours, as shown in Figure 2.

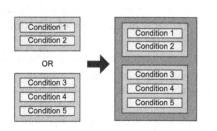
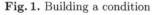
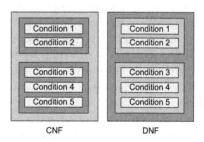

Fig. 1. Building a condition   **Fig. 2.** CNF vs. DNF

Actions can be sequenced or not. The former implies that the policy, when the condition is true, executes the actions in the order defined by the user. In this situation, as the user drops the actions in the specific place, the actions are connected with arrows.

We considered that the policy can be triggered by one or more events, so we defined a dark area to allow the user to drop a disjunction of events ($Event1 \lor Event2 \lor Event3$). The resulting interface groups roles, conditions, actions and events in a logical distribution and with well defined spatial distribution (Figure 3).

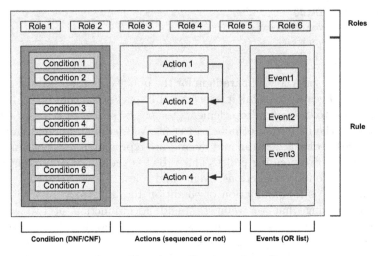

Fig. 3. Visual specification of a policy

4   Conclusions

The way users communicate with computational entities is usually based on some form of language. Textual languages, although easier to process by automatic processes, are seldom as user friendly as its visual equivalents. A visual sentence can be more attractive and easier to understand. In fact, humans recall faster, and understand better, a subject through some form of visual representation – it is common to use drawings to explain things.

We have defined a visual language for the definition of network management policies. This language works by dragging blocks, representing the actions, conditions and events, and dropping them in specific spaces for building a policy rule. This approach helps reducing the number of attributes to set as well as providing the user with visual aid in remembering the policy components. A well known analogy is the development of applications in a visual language, such as Visual Basic. Although less structured, it helps reducing the time-to-market and the rapid prototyping of applications when compared to textual languages, such as C++.

We hope that this approach helps reducing the difficulty in specifying and remembering policies which, some times, are written in a form difficult to understand and to remember.

References

1. Sloman, M.: Policy driven management for distributed systems. Journal of Network and Systems Management **2** (1994)
2. Westerinen, A., Schnizlein, J., Strassner, J., Scherling, M., Quinn, B., Herzog, S., Huynh, A., Carlson, M., Perry, J., Waldbusser, S.: Terminology for Policy-Based Management. RFC 3198, IETF (2001)
3. Stone, G., Lundy, B., Xie, G.: Network policy languages: A survey and a new approach. IEEE Network **15** (2001) 10–21
4. Damianou, N.: A Policy Framework for Management of Distributed Systems. PhD thesis, Imperial College (2002)
5. Shneiderman, B.: Direct manipulation. a step beyond programming languages. IEEE Transactions on Computers **16** (1983) 57–69
6. Wies, R.: Policies in network and system management – formal definition and architecture. Journal of Network and Systems Management **2** (1994)
7. Moore, B., Ellesson, E., Strassner, J., Westerinen, A.: Policy Core Information Model – Version 1 Specification. RFC 3060, IETF (2001)
8. Moore, B., Ed.: Policy Core Information Model (PCIM) Extensions. RFC 3460, IETF (2003)
9. Chomicki, J., Lobo, J., Naqvi, S.: Conflict resolution using logic programming. IEEE Transactions on Knowledge and Data Engineering **15** (2003)
10. Damianou, N., Dulay, N., Lupu, E., Sloman, M.: The ponder policy specification language. In: Policy 2001: Workshop on Policies for Distributed Systems and Networks, Bristol, UK, Springer-Verlag (2001)
11. OASIS: extensible access control markup language (xacml) version 2.0. Technical report, OASIS (2005)
12. Toktar, E., Jamhour, E., Maziero, C.: Rsvp policy control using xacml. In: IEEE 5th International Workshop on Policies for Distributed Systems and Networks (POLICY 2004), New York (2004)
13. Alaettinoglu, C., Villamizar, C., Gerich, E., Kessens, D., Meyer, D., Bates, T., Karrenberg, D., Terpstra, M.: Routing Policy Specification Language (RPSL). RFC 2622, IETF (1999)
14. Meyer, D., Schmitz, J., Orange, C., Prior, M., Alaettinoglu, C.: Using RPSL in Practice. RFC 2650, IETF (1999)
15. DMTF: Representation of cim in xml (xml mapping specification), v2.0.0. Technical report, DMTF (1999)
16. Strassner, J., Moore, B., Moats, R., Ellesson, E.: Policy Core Lightweight Directory Access Protocol (LDAP) Schema. RFC 3703, IETF (2004)
17. Lakin, F.: Spatial parsing for visual languages. In Chang, S.K., Ichikawa, T., Ligomenides, P., eds.: Visual Languages. Plenum Press, New York (1986) 35–85

Discovering Topologies at Router Level*

Donato Emma, Antonio Pescapé, and Giorgio Ventre

University of Napoli "Federico II"
{doemma,pescape,ventre}@unina.it

Abstract. Management is an essential task for the correct behavior of networks. In this field, several aspects should be taken into account. Among them, network topology is one of the most important elements to control. This paper proposes an approach to the topology discovery based on a hybrid methodology. We propose a tool, called *HyNeTD* (Hybrid Network-Topology Discovery), that effectively combines active and passive measurements to discover network topologies at router level. Architectural choices are presented and discussed and some preliminary experimental results, carried out over a controlled test-bed, are given.

1 Introduction

Automatic discovery of physical topology plays a crucial role in enhancing the manageability of modern IP networks. Discovering network topologies is an important and inherently difficult task [1] [3]. Network topology knowledge can prove useful in different situations, from *"fault isolation"*, to *"performance analysis"*, from *"network planning"* to *"service positioning"*, and, finally, from *"Traffic Engineering algorithms"* to a new general class of *"topology-aware distributed algorithms"*. Many factors contrast with the manual construction of network maps: (i) the number of entities involved in today's network is very large and keeps increasing exponentially; (ii) different parts of the network are managed by different authorities; (iii) network topologies are dynamic. One way to overcome these difficulties is to automate the discovery process by searching more efficient ways to find network maps. Several approaches for topology discovery at Autonomous System (AS) level have been proposed. As for approaches at router level some proposals exist yet, but, to the best of our knowledge, few platforms are available.

In this paper we propose an adaptive and hybrid IP based methodology for network-topology discovery at router level, its implementation in a tool called *HyNeTD* (Hybrid Network Topology Discovery), and its experimental validation.

The rest of the paper is organized as follows. Section 2 presents background and motivations at the base of our work. In Section 3 the state of the art is

---

* This work has been partially supported by the Italian Ministry for Education, University and Research (MIUR) in the framework of the FIRB Project "Middleware for advanced services over large-scale, wired-wireless distributed systems" (WEB-MINDS) and QUASAR PRIN Project. The authors would like to thank Domenico Dichiarante for his valuable support and hints

T. Magedanz, E.R.M. Madeira, and P. Dini (Eds.): IPOM 2005, LNCS 3751, pp. 118–129, 2005.

discussed. Section 4 presents *HyNeTD* design choices. In Section 5 a preliminary experimental analysis of our proposal is presented. Finally, Section 6 ends the paper with conclusions and issues for research.

2   Background and Motivations

An increasing number of Internet applications attempt to optimize their communications by monitoring network topologies. Hence, the need arises for widely usable, and highly accurate, algorithms and techniques capable to identify network entities using little or no information about them. Many topology discovery methodologies have been proposed in literature. We propose the following classification:

Passive Methodologies – relying on the use of Simple Network Management Protocol (SNMP) and Domain Name System (DNS);

Active Methodologies – based on the use of tools such *'ping'* and *'traceroute'*;

Routing Based Methodologies – based on the use of routing information;

Hybrid Methodologies – efficient combinations of the previous methodologies.

Different approaches to the topology discovery may be evaluated on the basis of *efficiency* (i.e. impose the least possible network overhead), *quickness* (i.e. take the least possible time), *completeness* (i.e. discover the entire topology) and *accuracy* (i.e. not make mistakes). *Passive Methodologies* are fast and reliable but also not always usable. *Active Methodologies* are neither reliable nor fast, but they are more widely usable. *Routing Based Methodologies* strictly depend on the routing protocol but they are fast and reliable. Finally, *Hybrid Methodologies* give the opportunity to merge the benefits of the previous methodologies. None of the above approaches outperforms the others. For example in [20] it is shown how studies based only on *traceroute* retrieved information may led to erroneous results. This consideration is pushing research in the topology discovery field towards an *adaptive hybrid methodology* able to follow the current network infrastructure status and configuration. In this paper we propose such an approach, by appropriately combining an active-based technique with a passive information retrieval methodology.

3   Related Work

Fremont system [12] uses an extensible set of discovery modules that are based on a variety of different protocols and information sources. To the best of our knowledge it was the first example of a hybrid discovery approach. In [2] Keshav *et al.* describe several heuristics and algorithms to discover both intra-domain and Internet-backbone topology. The proposed approach combines SNMP, routing information, *traceroute*, and measurements. Mansfield *et al.* in [11] define an SNMP based framework for *Internet mapping*. The main lack of their approach is the use of only passive information sources. In [9] Barford *et. al* study the marginal utility of adding information sources in performing wide-area measurements in the context of Internet topology discovery. They show that the marginal utility of additional measurement sites declines rapidly even after the first two sites. *Mercator* project [3] explores tools such as *traceroute* to group IP addresses in order to produce an Internet map. The CAIDA's Skitter [4] project has been developed to combine *traceroute* and benchmark-based analysis. This tool uses *traceroute* to find the paths connecting two nodes and to collect performance information from them. *Remos* [10] is an architecture that can be used to provide resources information to distributed applications. To collect information from different networks and their hosts, it provides several collectors that use different technologies, such as SNMP or benchmarking. Anyway, *Remos* gives support only to distributed applications, such for example grid computing, therefore it can not be used as a tool for *general purpose* network topology discovery. At AS level, there have been interesting works [5–8] which leverage *traceroute* measurements and BGP routes to infer AS-Level maps. In [13] Spring *et al.* present Internet mapping techniques that allow to measure directly router level ISP topologies without a significant loss in accuracy. The proposed techniques include the use of (i) BGP routing tables to focus measurements, (ii) IP routing properties , (iii) alias resolution techniques, and (iv) DNS queries. As for proprietary solutions, SNMP-based tools for automatic discovery of network topology are included in many commercial network management systems (i.e. HP's OpenView [14] and IBM's Tivoli [15]). These tools assume that SNMP is widely deployed, but they also send ICMP messages toward not SNMP-capable hosts and routers. Details of their discovery algorithms are proprietary and not available to the authors of this work.

4   HyNeTD Approach

HyNeTD main goal is to discover the router level topology of IP networks via the IP address spaces and the SNMP community names. *HyNeTD* has been designed to be used in scenarios that differ for (i) available information sources, (ii) size, and (iii) administrative policies. In any scenario *HyNeTD* tries to reach the maximum achievable accuracy and completeness, also reducing the produced network overhead.

In this Section we first introduce design choices adopted for *HyNeTD*, then we give a description of the *HyNeTD*'s architecture.

Design Choices. In order to guarantee a high degree of completeness and reliability, with a low network overhead, *HyNeTD* adopts a hybrid methodology by combining both passive and active approaches. SNMP and DNS are the information sources of the *HyNeTD*'s passive methodology. At the basis of the *HyNeTD*'s active approach are both *ping* with record route (*ping-rr* in the following) and *traceroute* utilities. *HyNeTD* applies the active approach only to get information from devices that do not have an accessible SNMP agent.

HyNeTD Architecture. *HyNeTD* implements a multi-step discovery process by adapting its behavior to the state of the network.

In the first step, in order to identify all the *active addresses*, *HyNeTD* sends a *ping* towards all the addresses belonging to the discovered addresses spaces. It assumes as active the responding ones. These addresses are added to a list of active addresses (the *UP list* in the following).

In the second step *HyNeTD* looks for available SNMP information sources. The simplest way to test the SNMP availability is to send an SNMP request and wait for a response. If a response arrives the SNMP availability may be assumed. This technique is simple but may produce a high network overhead if only a small part of the active machines have an accessible SNMP agent. In order to decrease this overhead, *HyNeTD* tries to reduce the number of addresses to be tested. It sends one UDP packet towards the *SNMP port* of any *active addresses*. *HyNeTD* assumes that it is impossible to obtain SNMP information from addresses responding with ICMP *"Host Unreachable-Port"* messages. Moreover, *HyNeTD* inserts such addresses in a list of addresses for which it is necessary to implement an active discovery process (the *ICMP list* in the following). If an address does not respond it is possible that (a) the UDP packet or the ICMP response is lost, (b) the device is configured to not respond, or (c) there is an active SNMP agent. In order to avoid mistakes due to (a), in case of no response *HyNeTD* resends the UDP packet. If still no response is received *HyNeTD* assumes (c) and inserts the address in the the list of addresses that may give SNMP information (the *SNMP list* in the following). Obviously (b) is still possible, *HyNeTD* solves this ambiguity at a later stage.

In the third step *HyNeTD* starts the passive discovery process. For each address belonging to the *SNMP list*, *HyNeTD* sends some SNMP requests to retrieve information belonging to the standard part of the SNMP Management Information Base (MIB) (see Table 1). *HyNeTD* assumes that all the addresses that do not respond to the first SNMP request can not be used as SNMP information sources. Therefore, it does not send any future SNMP request toward such addresses, which are also added to the *ICMP list*. Finally, If the DNS feature is enabled, *HyNeTD* ends the passive discovery process by sending DNS inverse look-up calls to obtain the addresses name.

In the forth step *HyNeTD* starts the active discovery process. This process is based on the use of both *ping-rr* and *traceroute*. In order to obtain information also from routers that do not respond to *traceroute*, *HyNeTD* uses a modified *traceroute* implementation: it uses ICMP Echo-Request messages instead of UDP packets with an invalid destination port. The information retrieved using *ping-rr*

Table 1. Retrieved MIB entries

IP-MIB::ipForwarding	an integer value equal to 1 if the device is capable to forward IP packets
IP-MIB::ipAdEntAddr	a table containing the ip addresses of the interfaces belonging to the device
IP-MIB::ipAdEntIfIndex	a table containing references to other interfaces from which it is possible to obtain SNMP information
IP-MIB::ipAdEntNetMask	a table containing the net masks of each of the device's addresses
IP-MIB::ipNetToMediaNetAddress	the ARP tables of each of the interfaces of the device
RFC1213-MIB::ipRouteDest	a table containing the subnets of all the interfaces reachable from the device
RFC1213-MIB::ipRouteNextHop	a table containing the addresses of the next hop routers
RFC1213-MIB::ipRouteMask	a table containing the net masks of all the reachable subnets
RFC1213-MIB::ipRouteType	a table containing for each of the subnets of the ipRouteMask a flag that indicates if the device is directly connected to the subnet
IF-MIB::ifSpeed	a table containing the speed of each of the interfaces belonging to the device

may be effective in discovering networks of small size. Indeed, by using *ping-rr* it is possible to discover some links that *traceroute* would have not found due to the presence of devices that do not respond to the ICMP Echo-Request.

Finally, during the fifth step, *HyNeTD* merges and elaborates all retrieved information. This process is composed of several sub-steps:

(a) *Passive subnets reconstruction.* Using information obtained from the SNMP, *HyNeTD* identifies the subnet of each address belonging to the *SNMP list*.

(b) *Passive links reconstruction.* *HyNeTD* identifies all the links between the routers discovered during the third step by using the following observation: "if two routers are connected through the interfaces A and B then those interfaces must belong to the same link".

(c) *Alias resolution.* During the fourth step *HyNeTD* has constructed several lists of addresses. In this sub-step it uses these lists to discover routers' addresses by using the following assumption: "if an address is on the path from the *HyNeTD* host to any destination, *HyNeTD* assumes that this address belongs to a router". Moreover, the discovered routers' addresses are not yet grouped together into routers. The process of grouping is called *alias resolution*. To overcame the drawbacks of a simple *alias resolution* approach based only on information retrieved from the DNS system (e.g. DNS may be mis-configured), *HyNeTD* combine such an approach with the *Ally alias-resolution* heuristic [13].

(d) *Active links reconstruction.* In this step *HyNeTD* uses the results of (c) in order to find out the links not recognized in (b). Starting from the same observation used in (b) and using the results of both *ping-rr* and *traceroute*, for each router's address A, *HyNeTD* searches among the addresses of the *previous hop devices* (devices that are one step before the considered router in either the result of *traceroute* or *ping-rr*) the one that is the *closer* to A (that is the address that produces the minimum value when XOR-ed with A). This address is assumed to be the other end of the link that starts from A.

(e) *Active subnets identification.* Finally, *HyNeTD* finds the subnet masks of each address belonging to the *ICMP list* performing the following operations:

- *Extraction of subnets from links:* For each link $A1 \longleftrightarrow A2$ identified during (d), *HyNeTD* calculates a temporary net-mask using the empirical formula $temp\_mask = NOT[(AND(A1, A2))XOR(OR(A1, A2))]$. $temp\_mask$ is an estimation of the unknown subnet mask. If the estimation is not compatible with $A1$ and $A2$, *HyNeTD* produces a compatible mask by setting to 0 the least significant bits of $temp\_mask$. The resulting mask is associated to both A1 and A2.

  The following sub-steps deal with addresses not associated to any link.

- *Passive information comparison:* HyNeTD verifies if it is possible to include each of these addresses into one of the already identified subnets and sets coherently their subnet masks.

- *Active addresses partitioning:* By assuming that two addresses belong to the same subnet if they have the same *previous hop address*, *HyNeTD* divides the addresses that still are not associated to any subnet in a set of lists. Each of these lists represents a subnet.

- *ICMP mask request:* *HyNeTD* sends an ICMP mask request towards all the addresses of each list. Each address may (a) respond with a valid mask, (b) respond with 0.0.0.0, or (c) not respond. If all the addresses of a list do not respond, or return an invalid mask, nothing can be done. In the case of just one valid response, *HyNeTD* assumes the given mask as the mask of the list. In case of multiple responses, it implements a *voting* process to select the list's net-mask.
- *Subnet guessing from a cluster of addresses:* For all still pending subnet lists, *HyNeTD* performs the *subnet guessing from a cluster of addresses* heuristic [2]. This heuristic returns an estimation of the net-masks to be assigned to the addresses lists.

HyNeTD Multi-thread Implementation. Both passive and active methods require to wait responses from contacted devices. *HyNeTD* tries to avoid these waits adopting a Multi-Thread Implementation in which all the activities that do not have a serial tie are overlapped: (i) the first four steps are characterized by an internal concurrent architecture; (ii) passive discovery methods and active discovery methods are executed in a parallel fashion.

To summarize, the proposed architecture, stepping from the results of some previous works, integrates different approaches and techniques and introduces some main innovations, such: (i) the use of *ping-rr*; (ii) the implementation of a scan technique to test the SNMP availability; (iii) the implementation of the Ally algorithm; (iv) and the use of *DNS inverse look-up* call in combination with the results of the Ally algorithm. These innovations, as shown in section 5, enable to achieve good performance in terms of efficiency, quickness, completeness, scalability and accuracy.

5   Validation and Experimental Analysis

The main goal of our experimental analysis has been to study the behavior of *HyNeTD* by varying, the discovered topologies (in this paper we present the result obtained analyzing the topologies of Figures 1, 2, and 3), the number of threads, the number of retries and the amount of "cross-traffic". The metrics used to evaluate *HyNeTD* are:

Network overhead – total amount of probing traffic;

Discovery time – discovery's duration;

Accuracy – ratio between the number of discovered routers R_d and the real number of routers.

Fig. 1. Ring topology   **Fig. 2.** Backup topology   **Fig. 3.** Linear topologies

5.1   Test-Bed Description

The used test-bed (see Table 2) is composed by 7 routers (Linux Mandrake 9.1 boxes with *ZEBRA Routing Software*), and by a Compaq Evo notebook with Linux RedHat 9.0 used to run *HyNeTD*. As for SNMP, we used the UCD-SNMP [16]. Moreover, we used D-ITG [19] to generate cross traffic.

Table 2. Test-bed description

Router/Host Names	CPU	RAM	Network Interfaces
CRONUS, POSEIDON, APHRODITE, ZEUS	Intel Celeron 500 MHz	64 MB	3 INTEL pro 100+
HELIOS, CALVIN, GAIA	Intel PIII 757 MHz	256 MB	3 INTEL pro 100+
HyNeTD host	Intel P IV 2.4 GHz	256 MB	NIC integrated on the board

Due to its adaptive and hybrid architecture, *HyNeTD* may be used in different network conditions. Here we show experimental results with *HyNeTD* used in the conditions reported in Table 3. In the *Passive* case *HyNeTD* uses only passive methods, so it works like a pure passive discovery tool. Vice versa, in the *Active* case it works like a pure active discovery tool. Clearly, in the *Hybrid* cases *HyNeTD* uses both passive and active methods. As for the comparison with existing approaches, thanks to this *modus operandi* we compared HyNeTD with pure passive and active approaches (implemented in other tools). Before stepping into experimental details, it is worth noting that we repeated each test several times. In the following the mean values across 20 test repetitions are reported and we obtained a confidence interval greater than or equal to 94%. Table 4 shows and describes all the *HyNeTD*'s input parameters and options. Its outputs are the lists of discovered *hosts*, *routers*, links, and *subnets*, provided both in text and XML format.

Table 3. Network Conditions

Name	Description
Passive	SNMP available on all the network routers. "DNS inverse look-up" feature enabled.
Active	SNMP disabled on all routers. "DNS inverse look-up" feature disabled.
Hybrid 1	SNMP disabled on all routers. "DNS inverse look-up" feature enabled.
Hybrid 2	SNMP available only on some network routers. "DNS inverse look-up" feature enabled.

Table 4. *HyNeTD*'s input parameters

Parameter's syntax	Parameter's description	Parameter's syntax	Parameter's description
-b base_address size [base_address size]	base addresses and dimensions of discovered address spaces	-d	enable DNS inverse name lookup
-c community_name [community_name]	SNMP's community names	-e	enable ARP table extraction
-r ret_number	max number of retries in packet sent	-n max_number_of_thread	maximum number of threads
-t time_out	timeout duration		

5.2   Ring Topology Experimental Results

Ring topology is characterized by a loop. Adopting an active methodology, and probing the network from a single point, the discovery of this loop fails. Indeed, in such a topology one router does not forward packets. In this case the topology can not be correctly reconstructed and such a router is not recognized as a router. This sub-section shows that *HyNeTD*'s hybrid approach allows to correctly reconstruct this loop using a limited amount of passively collected data.

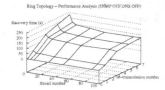

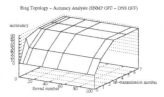

Fig. 4. Ring topology discovery time   **Fig. 5.** Ring topology accuracy

Discovery Time and Accuracy Evaluation. Figures 4 and 5 show the measured discovery time and accuracy trends in the *Active* case as function of the number of threads, and of the number of retransmissions. The discovery time is increasing with the number of retries and decreasing with the number of threads. The first dependence can be easily explained taking into account the extra waiting times due to each retransmission. The second dependence is related to a faster execution of the steps 1 and 2. There is no relation with the SNMP information retrieval. Indeed, the current implementation of *HyNeTD* does not parallelize the step 3. In all other network conditions we have found the same behaviour. Table 5 shows, for all network conditions, (i) the minimum and the maximum measured discovery time with and without cross traffic (column 2 and 3), (ii) the maximum theoretical and the achieved accuracy (column 4). Considering that we used a high amount of cross traffic ($100Mbps$), we measured a very light dependence on it. We found the same accuracy, a difference in the discovery time that can be neglected in the worst cases also varying the number of threads (see figure 6), and the same tendency for both these metrics. Therefore the following results represent well both the cases with and without cross traffic.

Table 5. Ring topology discovery time and accuracy

Network Conditions	No cross traffic	Cross traffic	Accuracy
Passive	min: 3.74s max: 141.10s	min: 9.64s max: 148.45s	theoretical: 100% max. measured: 100%
Active	min: 21.6s max: 232.023s	min: 23.64s max: 232.17s	theoretical: 86% max. measured: 86%
Hybrid 1	min: 8.5s max: 160.29s	min: 10.45s max: 160.33	theoretical: 86% max. measured: 86%
Hybrid 2	min: 9.64s max: 148.44s	min: 14.4s max: 175.72s	theoretical: 100% max. measured: 100%

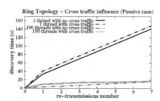

Fig. 6. Ring top. cross traffic analysis

As for the accuracy, we found the highest values in case of *Passive* and *Hybrid 2* conditions. In case of *Active* and *Hybrid 1* conditons the active methods are predominant, therefore there is a greater dependence on packets lost. *HyNeTD* considers that a packet is lost when *time_out* is elapsed. It is possible to accelerate the discovery process by reducing *time_out*, but some packets may be erroneously considered lost. In order to achieve the maximum theoretical accuracy in both *Active* and *Hybrid 1* conditions (that is equal to $\frac{R_d}{R_r} = \frac{6}{7} \simeq 0.86$) retries number was set equal to 3.

Network Overhead Analysis. The overhead produced by *HyNeTD* depends on several factors (i.e. network conditions, address spaces size, etc.). In this experimental analysis two address spaces were explored for a total amount of 90 addresses. Figure 7 shows sent and received data (in bytes) in all the network conditions. The *Passive* case presents the maximum overhead, whereas the lowest overhead is produced in case of *Hybrid 1*. This can be explained with a high amount of information retrieved from the DNS with a "low overhead". As for the produced overhead, considering the worst case (the *Passive* one), *HyNeTD* produces traffic with a mean rate that range from about $0.680KBps$ to about $10.4KBps$ (it depends on the maximum number of threads). Considering that the explored topology is composed of $100Mbps$ links this overhead may be neglected.

5.3   Backup Topology Experimental Results

Backup topology is characterized by the presence of a backup path. We considered this topology in order to verify if *HyNeTD* is capable to discover backup paths in case of network conditions different from the *Passive* one.

Discovery Time and Accuracy Evaluation. Table 6 shows that *HyNeTD* achives the minimum discovery time in the *Passive* case. In the *Hybrid 1* case it reaches similar performances. In the *Active* case it shows the worst discovery time. Finally, in the *Hybrid 2* case *HyNeTD* presents an intermediate discovery time. As for the accuracy, *HyNeTD* discovers correctly all routers, links, and subnets in all the considered network conditions. Such result shows that *HyNeTD* is capable to discover backup paths without using passive information sources.

Table 6. Backup topology discovery time and accuracy

Network Condition	No cross traffic	Accuracy
Passive	min: 6.27s max: 124.17s	theoretical: 100% max. measured: 100%
Active	min: 24.34s max: 204.44s	theoretical: 100% max. measured: 100%
Hybrid 1	min: 7.51s max: 129.4s	theoretical: 100% max. measured: 100%
Hybrid 2	min: 18.47s max: 159.17	theoretical: 100% max. measured: 100%

Network Overhead Analysis. The higher amount of overhead is produced in both *Passive* and *Active* network conditions (see Figure 8). The overhead produced in the *Active* case grows with the number of retries. Otherwise, in the *Passive* case it can be assumed independent on the number of retries. In the *Hybrid 1* network condition *HyNeTD* produces the minimum overhead. This result confirms the "high quality" of information retrieved using the DNS.

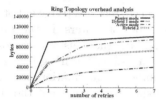

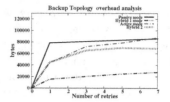

Fig. 7. Ring top. network overhead   **Fig. 8.** Backup top. network overhead

5.4   Linear Topologies Experimental Result

In this case we consider a family of *linear topologies* composed of a number of routers that ranges from 1 to 6. We used this family to evaluate the scalability of our tool. In testing how *HyNeTD* scales with respect to number of routers belonging to the discovered topology we considered especially the *Passive* and *Active* network conditions. Indeed, such conditions may be assumed respectively as the lower and the upper bound for the topology discovery times.

Discovery Time Analysis. Figure 9 shows the measured discovery times in case of *Passive* and *Active* conditions. The measured discovery time is quite independent from the number of routers in the *Passive* case, and highly dependent on it in the *Active* case. Hence, *HyNeTD* presents optimal scalability properties when used as a passive discovery tool. In the *Hybrid* cases the discovery time is always lower than the one measured in the *Active* case, and greater than the one measured in the *Passive* case. It is possible to state that *HyNeTD* is scalable when used as a hybrid discovery tool if in this case the discovery time is found to be close to the *Active* one. In order to verify if our hybrid approach is scalable, we considered the *Hybrid 2* network condition. Moreover, to consider a condition that is as much as possible close to the *Active* one we activated the SNMP agent on only one router. Figure 10 shows the discovery time in the case of *Active*, *Passive*, and *Hybrid 2* network conditions (values reported for the *Hybrid 2* are the worst ones measured varying the router on which the SNMP agent is active). In the *Hybrid* case, values are always less than one-half of that measured in the *Active* case. Moreover, they are very close to the *Passive* ones for the first three topologies, and their rates grow slower with respect to the *Active* case. Therefore, this preliminary analysis shows that *HyNeTD* hybrid approach scales better than a pure active approach.

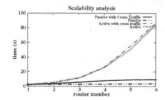

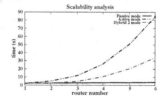

Fig. 9. Passive and Active cases   **Fig. 10.** Hybrid case

Network Overhead Analysis. As for the produced/received traffic in both the *Active* and *Passive* cases, it grows with the number of routers. In the *Passive* case we found a linear growing rate coherently with the constant size of the SNMP messages. In the *Active* case we found a "super-linear" growing rate coherently with the factorial growing of the number of information needed for the alias resolution process. Therefore, it is possible to assume that the *Active* mode is less scalable than the *Passive* one.

6   Conclusions and Future Works

In this paper we proposed a platform for the topology discovery of IP networks at router level. We presented the methodologies which our platform is based on as well as its performance analysis over classical network topologies. The proposed hybrid approach showed good performance in terms of accuracy, network overhead, and discovery time. Obtained results confirmed our hypothesis on the power of hybrid approaches. Indeed, the hybrid approaches are capable to adapt their behavior to the different network conditions. Moreover, due to the combined use of several techniques we showed how our hybrid approach permits to reach better performance overcoming the lacks of both active and passive methods. As for scalability, we found that our hybrid methodology presents an optimal behavior with respect to pure active approaches. Results presented in this paper are related to experiments over a controlled test-bed on a small-scale. Currently, we are testing the system behavior on realistic networks of a much wider-scale (i.e. Universitá di Napoli network). Preliminary results (both in terms of performance and scalability proprieties) confirming the behavior that HyNeTD has shown over a laboratory test-bed. We will extend our platform to discover "Layer 2" topologies too. We plan to adopt a hybrid methodology composed of several techniques [17] [18] able to take into account results from several data sources. Due to the low overhead of *HyNeTD* we plan to use it also in a distributed fashion, placing several monitoring probes over the network under control and partitioning the address space. The distributed monitoring platform will be enhanced with the use of tools we implemented for available bandwidth measurements [21]. Our objective is the development of a unified framework allowing to monitor and manage topology as well as links status.

References

1. C. Gkantsidis, E. Zegura, "Experiment and learn to discover Network Topology", Oct. 99 Georgia Tech.
2. R. Siamwalla, R. Sharma, and S. Keshav,"Discovering internet topology", Tech. Rep., Cornell University, Jul. 1998.
3. R. Govindan and H. Tangmunarunkit, "Heuristics for Internet map discovery", *Proc. of* IEEE Infocom '00, Mar. 2000.
4. K. C. Claffy and D. McRobb. "Measurement and Visualization of Internet Connectivity and Performance", http://www.caida.org/TOOLS/measurement/skitter/ (As of July 2005).

5. R. Govindan and A. Reddy, "An Analysis of Internet Inter-Domain Routing and Route Stability", *Proc. of* IEEE Infocom '97, Apr. 1997.
6. H. W. Braun and K. C. Claffy, "Global ISP interconnectivity by AS number", `http://moat.nlanr.net/AS/` (As of July 2005).
7. H. Chang, S. Jamin, and W. Willinger, "Inferring AS-level Internet Topology from Router-level Traceroutes", *Proc. of* SPIE ITCom '01, Scalability and Traffic Control in IP Networks, Aug. 2001.
8. H. Chang et al., "On inferring as-level connectivity from BGP routing tables", Technical Report UM-CSE-TR-454-02, 2002.
9. P. Barford et al., "The Marginal Utility of Network Topology Measurements". *Proc. of* ACM/SIGCOMM Internet Measurement Workshop, Nov. 2001.
10. P. Dinda et al., "The Architecture of the Remos System," *Proc. of* 10th IEEE Symp. on High-Perf. Dist. Comp. (HPDC'01), Aug. 2001.
11. G. Mansfield et al., "Techniques for automated Network Map Generation using SNMP". *Proc. of* Fifteenth Annual Joint Conference of the IEEE Computer and Communications Societies, Mar. 1996.
12. Wood et al., "Fremont: A System for Discovering Network Characteristics and Problems". *Proc. of* Winter USENIX Conference, Jan. 1993.
13. N. Spring, R. Mahajan, and D. Wetherall, "Measuring ISP Topologies with Rocketfuel". *Proc. of* ACM/SIGCOMM '02, Aug. 2002.
14. HP, "HP Management software", `http://www.openview.hp.com` (As of July 2005).
15. IBM, "Tivoli software", `http://www.tivoli.com` (As of July 2005).
16. University of California at Davis, "The UCD-SNMP project home page", `http://www.ece.ucdavis.edu/ucd-snmp/` (As of July 2005).
17. B. Lowekamp et al., "Topology Discovery for Large Ethernet Networks". *Proc. of* ACM SIGCOMM '01, Aug. 2001.
18. R. Black et al., "Ethernet Topology Discovery without Network Assistance", *Proc. of* 12th IEEE International Conference on Network Protocols (ICNP'04), Oct. 2004.
19. S. Avallone et al., "Performance evaluation of an open distributed platform for realistic traffic generation", Performance Evaluation: An International Journal, Vol. 60 issue 1/4, pp. 359-392, Mar. 2005
20. A. Lakhina, J. Byers, M. Crovella and P. Xie, "Sampling Biases in IP Topology Measurements", *Proc. of* IEEE Infocom, Apr. 2003.
21. A. Botta et al., "BET: A Hybrid Bandwidth Estimation Tool", *Proc. of* IEEE ICPADS, PMAC-PDG'05 Workshop, Vol. 2, pp. 520-524, Jul. 2005.

Comprehensive Solution for Anomaly-Free BGP

Ravi Musunuri and Jorge A. Cobb

Department of Computer Science, The University of Texas at Dallas,
Richardson, TX-75083-0688
{musunuri,cobb}@utdallas.edu

Abstract. The Internet consists of many self-administered and inter-connected Autonomous Systems (ASms). ASms exchange inter-AS routing information with each other via the Border Gateway Protocol (BGP). Neighboring BGP routers located in different ASms share their inter-AS routing information via external BGP (eBGP), whereas two routers in the same AS share their inter-AS routing information via internal BGP (iBGP).

From the paths received from its peers, each BGP router chooses the best path based on routing policies chosen locally at its own AS. Conflicting policies between different ASms may cause divergence problems in eBGP, i.e., permanent oscillations in the chosen path to the destination. On the other hand, divergence problems may also occur in iBGP. This is caused by the interaction of route-reflection clustering, which is a technique to improve the scalability of iBGP, and other factors, such as intra-AS link costs, among others. In this paper, we provide a comprehensive solution that avoids all the known anomalies with both eBGP and iBGP. In our solution, each AS can locally choose its routing policies, while still ensuring anomaly-free behavior.

1   Introduction

The Internet consists of many self-administered and inter-connected Autonomous Systems (ASms). Routing in the Internet is separated into intra-AS routing and inter-AS routing. Intra-AS routing (e.g., OSPF, RIP) advertises routing information that is local to the AS to all routers within the same AS. The Border Gateway Protocol (BGP) [1] advertises inter-AS routing information between BGP routers. Each pair of neighboring BGP routers in different ASms share their inter-AS routing information via external BGP (eBGP). Each pair of BGP routers within the same AS share their inter-AS routing information via internal BGP (iBGP). Contrary to eBGP, sharing routing information in iBGP is done even if the pair of routers are not neighbors, i.e., even if they are separated by multiple network hops.

BGP routers exchange inter-AS routing information via a TCP connection with each of its BGP peers. If a peer is located in a different AS, it is known as an eBGP peer, and the TCP connection to this peer is referred as an eBGP peering session. Similarly, a peer in the same AS is known as an iBGP peer, and the TCP connection to it is referred as an iBGP peering session.

From the set of paths advertised by its peers, each router chooses the best path based on the routing policies chosen locally at its AS. Conflicting routing policies [2] between different ASms may cause divergence in eBGP, i.e., the path to the destination

T. Magedanz, E.R.M. Madeira, and P. Dini (Eds.): IPOM 2005, LNCS 3751, pp. 130–141, 2005.

continuously oscillates between several possible paths. Divergence may also occur in iBGP, even when eBGP is stable. This is caused by the interaction of route-reflection clustering [3], which is a technique to improve the scalability of iBGP, and other factors, such as intra-AS link costs, and MED values[1].

Many solutions have been proposed to avoid eBGP and iBGP divergence anomalies separately. However, we are not aware of any solution that solves divergence anomalies in both eBGP and iBGP combined. Govindan et al. [4] proposed an architecture to analyze the routing policies statically and find conflicting routing policies. On the other hand, Griffin et al. [5] have shown that checking routing policies for divergence is intractable. Gao et al. [6] proposed a set of guidelines for choosing routing policies in order to avoid eBGP divergence. However, their solution removes the freedom of each AS to choose its routing policies locally. Basu et al. [7] and Walton et al. [8] provided solutions to solve iBGP divergence anomalies. For a given destination prefix, in the original iBGP, each router only advertises a single best path to its iBGP peers. In both of these solutions, for a given destination, routers are required to advertise multiple paths to its iBGP peers, requiring higher memory and message overheads, and thus defeating the purpose of using route-reflection clustering.

In this paper, we are providing a comprehensive solution that solves all the known anomalies with iBGP and eBGP. In our solution, BGP path update message carries only two additional integer cost metric values. One cost metric is used to detect and avoid the eBGP divergence anomalies and other cost metric is used to detect and avoid iBGP divergence. For a given destination prefix, our solution does not require multiple path advertisements between iBGP peers. Also, each AS can choose routing policies locally. Our solution restricts the routing policies, only when, there exists a divergence.

2 BGP Path Selection

In this paper, we assume that each router tries to find a path to some special destination prefix d. A path P received by a router R located in AS u contains the following attributes:

- $local\_pref$: A preference value indicating the ranking of P in the local routing policy of AS u. A larger preference value indicates a greater preference for the path.
- AS_path: Sequence of ASms along the path to reach the destination prefix d from the current AS u.
- MED: For a pair of ASms connected by more than one link, the Multi-Exit Discriminator (MED) value indicates the preference of one link over another. A smaller MED value indicates a greater link preference.
- $next\_hop$: The IP address of the next-hop border router. If the router R is an interior router then $next\_hop$ is the IP address of the border router that is the exit point from u. If the router R is a border router then $next\_hop$ is the IP address of the border router that is the entry point into the neighboring AS.

---

[1] The Multi-Exit-Discriminator (MED) value is used to define the preference level of inter-AS links when the pair of neighboring ASms are connected by more than one inter-AS link

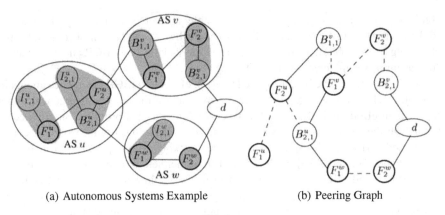

(a) Autonomous Systems Example   (b) Peering Graph

Fig. 1.

---

best(input A: set of paths advertised by peers)
{

1. A is reduced to only those paths with largest $local\_pref$ value.
2. If $|A| > 1$, then reduce A to those paths with least AS_path sequence length.
3. If $|A| > 1$, then separate A into disjoint subsets, where all paths in a subset exit via the same neighboring AS. Reduce each subset to those paths with smallest MED value. Set A to the union of the reduced subsets.
4. If $|A| > 1$, then:
    (a) If A has at least one path whose $next\_hop$ is an eBGP peer, then the router reduces A to those paths whose $next\_hop$ is an eBGP peer.
    (b) If A has no paths whose $next\_hop$ is an eBGP peer, then the router reduces A to those paths whose intra-AS cost from itself to the path's border router is the least.
5. Finally, if $|A| > 1$, then use some deterministic tie breaker to reduce A to a single element.
6. The best path is the single element in A.

}

Fig. 2. Best Path Selection Algorithm

From each peer, a router receives a path (potentially empty) to reach the destination. From this set of paths, the router must choose the "best" path and adopt it as its own path. The best path is chosen according to the algorithm given in Fig. 2 [7]. If a router adopts a new path, i.e. if its best path is not its previously chosen path, then the router informs each of its peers about the newly chosen path.

3   Route-Reflection Clustering

In the original iBGP peering scheme, each border router within an AS is a iBGP peer of all other routers within the same AS. As the size of the AS increases, this scheme fails to scale due to large number of iBGP peering sessions required. A common solution is to employ route-reflection clustering [3]. In this approach, the routers within an AS are divided into disjoint sets, known as *clusters*. In Fig. 1(a), AS u is divided into two clusters depicted by the shaded regions. One distinguished router in each cluster is known as the *reflector*. The reflector within AS u and cluster i is denoted F_i^u, and to highlight this node, it is drawn in bold. Border routers within AS u and cluster i are denoted by $B_{i,j}^u$ for some j, and likewise interior routers within AS u and cluster i are denoted by $I_{i,j}^u$ for some j.

Each reflector maintains a peering session with routers that fall in the following three categories: (a) all routers within its own cluster (via iBGP peering), (b) all reflectors of all other clusters in its AS (via iBGP peering), (c) in the case when the reflector is also a border router, all its neighboring routers outside of its AS (via eBGP peering). All routers, within its cluster, that establish a iBGP peering session with a reflector are known as the *clients* of the reflector. For example, in Fig. 1(a), the clients of reflector F_2^u are $I_{2,1}^u$ and $B_{2,1}^u$.

Note that interior routers learn about paths to the destination only via their reflector. Furthermore, although border routers may learn paths from their neighbors outside of their AS, the only router within their own AS from whom they learn paths is their reflector. As an example, consider again Fig. 1(a), in particular, border router $B_{2,1}^u$. Although it has a eBGP peering session with its neighbor in AS v and learns paths from it, the only router within its own AS u from whom it may learn a path is its reflector F_2^u. In particular, notice that even though $B_{2,1}^u$ is a neighbor of both F_1^u and $I_{2,1}^u$, it does not establish a peering session with these routers.

Reflector, F_i^u, advertises its best path to other peers as explained below:

– If F_i^u received its best path from another reflector, then F_i^u advertises its best path to all its clients and eBGP peers.
– If F_i^u received its best path from a client or from an eBGP peer, then F_i^u advertises its best path to all reflectors, to all its clients, and to all its eBGP peers (except the router from whom the best path was received).

4   Greedy Protocol

In this section, we will reduce the BGP routing problem into an abstract and formal notation known as the Stable Paths Problem (SPP). The SPP was originally introduced by Griffin et.al. [5] to model eBGP routing. However, in [10] and [12] it was shown that the SPP model can be extended to model iBGP routing. In this paper, we will extend the SPP model to model eBGP and iBGP combined.

An SPP instance consists of a tuple $(G, \prec)$, where G is a graph and $\prec$ is a ranking relation between the paths along the graph G.

For our purposes, we restrict G as follows. Each node in G corresponds to either a border router or a reflector router and each edge corresponds to a peering session between two routers. Notice that interior and non-reflector routers are removed from the peering graph. In general, interior and non-reflector routers do not effect the path selection; they only choose the path advertised by their reflector. Figure 1(b) presents the peering graph of the example shown in Fig. 1(a). eBGP peering sessions are shown as solid lines and iBGP peering sessions are shown as dotted lines.

Next, we define $\prec$, which is a ranking relation between the paths at a node along the peering graph G. We define $P \prec Q$ at node x, where both paths P and Q originate at node x and end at node d, as follows.

$$P \prec Q \equiv ((Q = best(\{P, Q\})) \wedge P \neq Q)$$

I.e., router x prefers Q over P when these are its only available choices.

We require relation $\prec$ to be a total-order on paths. However, a total-order is not guaranteed if the best-path selection algorithm uses MED values. Until section 7, we will ignore the MED values for path selection. In section 8, we will briefly discuss incorporating MED values into our approach. In this paper, we will use eBGP results from [11] and iBGP results from [12] to provide a comprehensive solution.

Every node x chooses a path to d among the paths offered by its neighbors in the peering graph. The path currently chosen by x is denoted by $\pi(x)$. This path is a sequence of nodes represented as $\langle x\ y \ldots d \rangle$, and its value is updated under the following constraints.

- At all times, $\pi(x)$ should be a loop-free path or the empty path.
- Node x can update $\pi(x)$ only by assigning to it the path $\langle x\ \pi(y) \rangle$ for some neighbor y in the peering graph.

Note that the path actually taken by datagrams as they traverse an AS may be different from the chosen paths above. E.g., if $\pi(x)$ is equal to $\langle x \ldots u \ldots v \ldots d \rangle$, where all the nodes in the sub-path $\langle u \ldots v \rangle$ belong to the same AS, the actual path taken by datagrams is the shortest intra-AS path between routers u and v, which may be different from the sub-path from u to v in $\pi(x)$.

Every node receives at most one path from each its peers. The set of paths advertised by all the peers of node x is denoted by $choices(x)$.

Next, we will present the greedy protocol, which simulates the working of BGP protocol with route-reflection. Specification of the greedy protocol at node x is shown in Fig. 3. The notation used in this paper is similar to the notation defined in [13], [14]. The greedy protocol consists of one action with guard $\pi(x) \neq best(choices(x))$. If the guard is true, i.e., if the current chosen path is different from the best available path, then node x greedily assigns to $\pi(x)$ the path $best(choices(x))$.

```
node x
begin
    π(x) ≠ best(choices(x)) → π(x) := best(choices(x))
end
```

Fig. 3. Greedy Protocol

In the next two sections, we present two anomalies associated with the greedy protocol.

5   eBGP Divergence

Since the rank of each path is chosen arbitrarily at each AS, conflicting choices at neighboring ASms may prevent ASms from maintaining a stable path. That is, paths chosen by some ASms may oscillate continuously, even though neither G nor $\prec$ change.

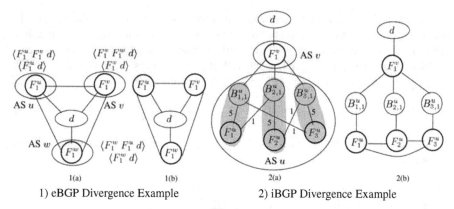

Fig. 4.

1) eBGP Divergence Example   2) iBGP Divergence Example

Consider the SPP instance shown in Fig. 4.1(a)[2]. Each AS contains only one BGP router. Figure 4.1(b) presents the peering graph of the SPP instance shown in Fig. 4.1(a). The paths acceptable to an AS (i.e. ranked higher than the empty path) are alongside the AS in the decreasing order of rank. Note that each AS prefers longer paths over shorter paths. E.g., F_1^u prefers the longer path $\langle F_1^u, F_1^v, d \rangle$ over the shorter path $\langle F_1^u, d \rangle$. This causes the ranking of each node to be in conflict with the ranking of its next hop to d.

The cyclic relationship between these ranking prevents any node from obtaining a stable path to d. To see this, consider the following steps:

- Initially F_1^u, F_1^v, and F_1^w choose the paths $\langle F_1^u, F_1^v, d \rangle$, $\langle F_1^v, d \rangle$, and $\langle F_1^w, d \rangle$, respectively.
- Node F_1^v notices that F_1^w chose the path $\langle F_1^w, d \rangle$. Hence, F_1^v changes its path to $\langle F_1^v, F_1^w, d \rangle$. This in turn forces F_1^u to change its path to $\langle F_1^u, d \rangle$.
- Node F_1^w notices that F_1^u chose the path $\langle F_1^u, d \rangle$. Hence, F_1^w changes its path to $\langle F_1^w, F_1^u, d \rangle$. This in turn forces F_1^v to change its path to $\langle F_1^v, d \rangle$.
- Node F_1^u notices that F_1^v chose the path $\langle F_1^v, d \rangle$. Hence, F_1^u changes its path to $\langle F_1^u, F_1^v, d \rangle$. This in turn forces node F_1^w to change its path to $\langle F_1^w, d \rangle$, and the system is back to its initial state.

Converging to a steady state is highly sensitive to the ranking of paths. For instance, in Fig. 4.1, if the ranking of paths at F_1^u is reversed, then the system is guaranteed to converge to a steady state. Due to this sensitivity to the ranking of paths, deciding if an SPP instance converges is NP-complete [5].

6   iBGP Divergence

Even with stable eBGP, we consider an iBGP anomaly in which routers within an AS fail to converge to a stable assignment of paths [9]. We refer to this anomaly as clustering-induced divergence, because the interaction between route-reflection clustering and intra-AS routing link costs causes the system to diverge.

---

[2] This SPP instance is known as BAD GADGET in [2], [5]

An example of clustering-induced divergence is shown in Fig. 4.2(a) [9]. Figure 4.2(b) shows the peering graph of Fig. 4.2(a). In this example, we assume that at AS u, local preference values of all the available paths to destination prefix d are equal.

Note that in the peer graph, each reflector F_i^u always prefers path $\langle F_i^u, F_{i+1}^u, B_{(i+1,1)}^u, F_1^v, d \rangle$ over path $\langle F_i^u, B_{i,1}^u, F_1^v, d \rangle$ due to following[3]:

$$cost(F_i^u, B_{i,1}^u) > cost(F_i^u, B_{(i+1,1)}^u). \tag{1}$$

Initially, assume that for each i, $\pi(F_i^u) = \langle F_i^u, B_{i,1}^u, F_1^v, d \rangle$. Consider the following sequence of events.

1. F_1 changes its path to $\pi(F_1^u) = \langle F_1^u, F_2^u, B_{2,1}^u, F_1^v, d \rangle$ because the path via F_2^u is ranked higher than its current path via $B_{1,1}^u$.
2. F_2^u changes its path to $\pi(F_2^u) = \langle F_2^u, F_3^u, B_{3,1}^u, F_1^v, d \rangle$ because the path via F_3^u is ranked higher than its current path via $B_{2,1}^u$.
3. F_1^u returns its path to $\pi(F_1^u) = \langle F_1^u, B_{1,1}^u, F_1^v, d \rangle$, because its previous path via F_2^u is no longer available.
4. F_3^u changes its path to $\pi(F_3^u) = \langle F_3^u, F_1^u, B_{1,1}^u, F_1^v, d \rangle$, because the path via F_1^u is ranked higher than its current path via $B_{3,1}^u$.
5. F_2^u returns its path to $\pi(F_2^u) = \langle F_2^u, B_{2,1}^u, F_1^v, d \rangle$, because its previous path via F_3 is no longer available.
6. F_1^u changes its path to $\pi(F_1^u) = \langle F_1^u, F_2^u, B_{2,1}^u, F_1^v, d \rangle$ because the path via F_2^u is ranked higher than its current path via $B_{1,1}^u$.
7. F_3^u returns its path to $\pi(F_3^u) = \langle F_3^u, B_{3,1}^u, F_1^v, d \rangle$, because its previous path via F_1^u is no longer available.

The state of the system after step 7 is the same as the state after step 1. The system will therefore never reach a steady assignments of paths.

7   Comprehensive Solution

To provide a comprehensive solution that solves divergence anomalies with both eBGP and iBGP, we combine the solutions provided in [11], [12]. The general behavior of both solutions is similar. However, they have significant differences. The eBGP solution [11] models each AS as a single node. On the other hand, the iBGP solution [12] does consider the individual routers within a single AS, but it assumes the external paths from border routers are stable.

In both the solutions, each node maintains a cost value to detect divergence. Cost values grow without bound if there exists divergence in the system. If the cost value grows above some threshold value C, then nodes restrict their routing policies such that divergence is removed from the system. For our comprehensive solution, we have to decide whether to have two separate cost values to solve each of eBGP and iBGP divergence anomalies, or to have a single cost value to solve both eBGP and iBGP divergence anomalies.

---

[3] Note that mod 3 is implied on the subscript i

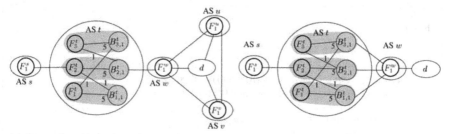

(a) Example with both eBGP and iBGP divergence

(b) Example with iBGP divergence

Fig. 5.

A simple and efficient solution could be to just maintain a single cost value for solving both eBGP and iBGP divergence anomalies. Figure 5(a) shows an example with both eBGP and iBGP divergence by combining Fig. 4.1(a) and Fig. 4.2(a). In this example, eventually, the cost at a node F_1^u or F_1^v or F_1^w reaches a maximum value of C due to eBGP divergence between ASms u, v, and w. This causes the node to stop changing its chosen path. Thus, the path chosen at this node becomes stable, which stabilizes the paths at all other nodes. Hence, the single cost metric solution solves divergence anomalies with both eBGP and iBGP. However, it is not practical due to following reason.

Lets consider Fig. 5(b) obtained by removing AS u and AS v nodes from Fig. 5(a). In this example, eBGP is stable, where as iBGP in AS t diverges. Due to divergence in AS t, the cost value at router F_2^t will increase continuously. If we use a single cost metric solution, then iBGP divergence in one AS could effect many ASms. For example, divergence in AS t is effecting AS s, even though the divergence is internal to t.

To avoid these unnecessary effects, we have to use two separate cost metric values to solve each of eBGP and iBGP divergence anomalies. Thus, we denote the eBGP cost at node x by $cost_e(x)$ and the iBGP cost at node x by $cost_i(x)$. Next, we will present a Comprehensive-Divergence Avoidance Protocol, which forces convergence in both eBGP and iBGP.

7.1 Comprehensive-Divergence Avoidance Protocol

Figure 6 shows the specification of the Comprehensive-Divergence Avoidance Protocol (C-DAP). By ignoring MED values, the ranking relation $\prec$ becomes a total-order relation at each node. This protocol is motivated by the following observations. Lets consider eBGP divergence steps shown in the Fig. 4.1(a). Rank of the best path at each node decreases periodically. Divergence may not be possible, if the rank of the best path at each node increases monotonically. Eventually, every node should get the highest ranked path as the best path and the system should stabilize. We use this observation to detect divergence in eBGP. We use similar observation to detect iBGP divergence. In iBGP divergence example, rank of the best path at each reflector decreases periodically. In particular, the rank of the best path at reflector F_1^u decreases from step 1 to step 3.

node x
begin

$\pi(x) = \langle u, \pi(par(x)) \rangle \rightarrow$
  $cost_i(x) := 0$ **if** $par(x) \in E$;
  $cost_i(x) := max(cost_i(x), cost_i(par(x)))$ **if** $par(x) \in I$;
  $cost_e(x) := max(cost_e(x), cost_e(par(x)))$;

[]
$\pi(x) \succ best(choices(x)) \wedge par(x) \neq newpar(x) \rightarrow$
  $\pi(x) := best(choices(x))$;
  $cost_i(x) := 0$ **if** $par(x) \in E$;
  $cost_i(x) := cost_i(x) + 1$ **if** $par(x) \in I$;
  $cost_e(x) := cost_e(x) + 1$ **if** $par(x) \in E$;
  $cost_e(x) := cost_e(par(x))$ **if** $par(x) \in I$;

[]
$\pi(x) \succ best(choices(x)) \wedge par(x) = newpar(x) \rightarrow$
  $\pi(x) := best(choices(x))$;
  $cost_i(x) := 0$ **if** $par(x) \in E$;
  $cost_i(x) := cost_i(par(x))$ **if** $par(x) \in I$;
  $cost_e(x) := cost_e(par(x))$;

[]
$\pi(x) \prec best(choices(x)) \wedge newpar(x) \in I \wedge (cost_i(newpar(x)) < C_i) \vee \pi(x) = \langle\rangle) \rightarrow$
  $\pi(x) := best(choices(x))$;
  $cost_i(x) := cost_i(par(x))$;
  $cost_e(x) := cost_e(par(x))$;

[]
$\pi(x) \prec best(choices(x)) \wedge newpar(x) \in E \wedge (cost_e(newpar(x)) < C_e) \vee \pi(x) = \langle\rangle) \rightarrow$
  $\pi(x) := best(choices(x))$;
  $cost_i(x) := 0$;
  $cost_e(x) := cost_e(par(x))$;

end

Fig. 6. Comprehensive-Divergence Avoidance Protocol

In C-DAP, every node x maintains a pair cost values, $cost_e(x)$ and $cost_i(x)$. eBGP cost value, $cost_e(x)$, is increased by one, whenever the rank of the best path decreases and the best path is advertised by an eBGP peer. Similarly, iBGP cost value, $cost_i(x)$, is increased by one, whenever the rank of the best path decreases and the best path is advertised by an iBGP peer. Whenever the paths advertised between eBGP peers are oscillating continuously, value of $cost_e$ metric increases without bound. If $cost_e$ increases beyond some threshold value C_e then the node infers that the eBGP divergence is occurring. Each node takes a corrective action to remove oscillations in eBGP paths. Similarly, in an AS, value of local $cost_i$ metric increases without bound, if the paths advertised between iBGP peers are oscillating. If $cost_i$ increases beyond some threshold value C_i then the node infers that the iBGP divergence is occurring. Each node takes a corrective action to remove oscillations in iBGP paths.

Before explaining the protocol in detail, we will discuss some notation used in presenting the protocol. For every node x, $par(x)$ denotes the next hop node along the current path, $\pi(x)$, and $newpar(x)$ denotes the next hop node along the new best path,

$best(choices(x))$. At node x, E represents the set of eBGP peer identifiers and I represents the set of iBGP peer identifiers. C-DAP specification consists of five actions. Each action updates $\pi(x)$, $cost_i(x)$, and $cost_e(x)$ variables depending on the guard condition, on which, they were excuted.

First action is enabled, if node x receives an update message from the current next-hop node, $par(x)$, with the current path, $\pi(x)$. This action consists of three assignment statements. If $par(x)$ is an eBGP peer then the first statement assigns $cost_i(x)$ variable to zero. Whenever a node x receives a path from an eBGP peer, node x resets the iBGP cost value to zero. If $par(x)$ is an iBGP peer then the second statement assigns $cost_i(x)$ variable to the maximum value of $cost_i$ at node x and $cost_i$ at $par(x)$. In the third statement, $cost_e(x)$ variable is always assigned to the maximum value of $cost_e$ at node x and $cost_e$ at $par(x)$.

Second action is enabled, if the rank of $best(choices(x))$ is lesser than the rank of $\pi(x)$ and $par(x)$ is not same as $newpar(x)$. This action consists of five assignments statements. First statement assigns $\pi(x)$ to the best path, $best(choices(x))$. After excution of the first statement, $par(x)$ and $newpar(x)$ are same. Next two statements update the $cost_i$ value based on whether $par(x)$ belongs to E or I. If $par(x)$ belongs to E then $cost_i$ is assigned to zero. If $par(x)$ belongs to I then $cost_i$ is increased by one. Last two statements update the $cost_e$ value depending on whether $par(x)$ belongs to E or I. If $par(x)$ belongs to E then node x increases the $cost_e$ value by one. But if $par(x)$ belongs to I then node x assigns its $cost_e$ value to $cost_e$ at $par(x)$.

Third action is enabled, if the rank of $best(choices(x))$ is lesser than the rank of $\pi(x)$ and $par(x)$ is equal to $newpar(x)$. This action consists of four assigment statements. First statement assigns $\pi(x)$ to the best path, $best(choices(x))$. Next two assignments update the $cost_i$ value based on whether $par(x)$ belongs to E or I. If $par(x)$ belongs to E then $cost_i$ is assigned to zero. If $par(x)$ belongs to I then $cost_i$ is assigned to $cost_i$ at $par(x)$. Value of $cost_e$ is always assigned to $cost_e$ at $par(x)$.

Fourth action is enabled, if the rank of $best(choices(x))$ is greater than the rank of $\pi(x)$, $newpar(x)$ belongs to I, and either $cost_i$ at $newpar(x)$ is less than some threshold constant C_i or $\pi(x)$ is empty. This action consits of three assignment statements. First statement assigns $\pi(x)$ to $best(choices(x))$. Next two statements assign $cost_i$, $cost_e$ values to corresponding cost values at $par(x)$.

Fifth action is enabled, if the rank of $best(choices(x))$ is greater than the rank of $\pi(x)$, $newpar(x)$ belongs to E, and either $cost_e$ at $newpar(x)$ is less than some threshold constant C_e or $\pi(x)$ is empty. First statement assigns $\pi(x)$ to $best(choices(x))$. Second statement assigns $cost_i$ value to zero and third statement assigns $cost_e$ value to $cost_e$ value at $par(x)$.

8   Other iBGP Anomalies

iBGP also suffers from two other types anomalies: MED-induced divergence, clustering-induced loops. Clustering-induced loops occur due to interaction between intra-AS link costs and route-reflection clustering. This anomaly can be avoided if the reflectors selectively advertise paths to their client routers. For complete details of this anomaly and proposed solution, readers are referred to [15]. iBGP also suffers from MED-Induced divergence anomaly. This anomaly disappears, if we ignore MED values for route selection. Reason for this anomaly is due to interaction between intra-AS routing link costs, route-reflection clustering, and MED values. MED-induced anomaly can be avoided by introducing virtual nodes [10], [12] in peering graph. Due space constaints, we are not presenting the complete details.

9   Related Work

There are several solutions proposed to solve eBGP and iBGP anomalies separately. We are not aware of any proposed comprehensive solution, that avoids anomalies with both eBGP and iBGP.

Proposed eBGP solutions can be divided into three categories. First category of solutions avoid the eBGP divergence by statically checking for conflicting routing policies in a centralized database [4]. This solution has several disadvantages. First, it requires global-coordination among all ASms. Second, Griffin et al. [5] also proved that the checking of conflicting routing policies is NP-hard. Second category of solutions avoid the divergence by presenting guidelines [6] for selecting the routing policies at each AS. This solution does not require global-coordination among ASms. But, it restricts the routing policies and removes the freedom of each AS choosing routing policies locally. Third category of solutions [16] avoid the divergence by restricting routing policies during runtime. In [16], every path update message carries the history of path update events. If a node finds a loop in the history of path update events then it removes some valid path(s) to avoid divergence. Loop in the history is only a necessary but not sufficient condition for divergence. Hence, their solution, sometimes, removes the path unnecessarily.

Proposed iBGP solutions avoid divergence by advertising multiple paths [7] [8] between each pair iBGP peers. Both solutions require high memory and message overheads. This defeats the whole purpose of using route-reflection clustering.

10   Summary and Concluding Remarks

BGP is the de-facto standard for inter-AS routing. Both external and internal forms of BGP plagued with many forms of anomalies. In this paper, we provided a comprehensive solution that solves all the known anomalies. Specification of our C-DAP protocol assumes shared memory model. But, we can easily change to more general message passing model by assuming that each path update message carries a pair of integer cost values. In our protocol, divergence increases the cost values to the maximum threshold values. We can reset these cost values by maintaining timers or by periodically using a reset protocol presented in [17].

References

1. Y. Rekhter and T. Li, "A border gateway protocol," *IETF RFC-1771*, 1995.
2. T. G. Griffin, F. B. Shepherd, and G. Wilfong, "Policy disputes in path vector protocols," in *Proc. of IEEE ICNP conference*, 1999, pp. 21–30.

3. T. Bates and R. Chandrasekeran, "BGP route reflection - an alternative to full-mesh IBGP," *IETF RFC-1966*, 1996.

4. R. Govindan, C. Alaettinoglu, G. Eddy, D. Kessens, S. Kumar, and W. S. Lee, "An architecture for stable, analyzable Internet routing," *IEEE Network*, vol. 13, no. 1, pp. 29–35, 1999.

5. T. G. Griffin, F. B. Shepherd, and G. Wilfong, "The stable paths problem and interdomain routing," *IEEE/ACM Trans. Networking*, vol. 10, no. 2, pp. 232–243, 2002.

6. L. Gao and J. Rexford, "Stable Internet routing without global coordination," *IEEE/ACM Trans. Networking*, vol. 9, no. 6, pp. 681–692, 2001.

7. A. Basu, C.-H. L. Ong, A. Rasala, F. B. Shepherd, and G. Wilfong, "Route oscillations in IBGP with route reflection," in *Proc. of ACM SIGCOMM conference*, 2002, pp. 235–247.

8. D. Walton, D. Cook, A. Retana, and J. Scudder, "BGP persistent route oscillation solution," *IETF Internet Draft*, 2002.

9. T. G. Griffin and G. Wilfong, "On the correctness of IBGP configuration," in *Proc. of ACM SIGCOMM conference*, 2002, pp. 17–29.

10. ——, "Analysis of the MED oscillation problem in BGP," in *Proc. of IEEE ICNP conference*, 2002, pp. 90–99.

11. J. A. Cobb and R. Musunuri, "Convergence of inter-domain routing," in *Proc. of IEEE GLOBECOM conference*, 2004, pp. 1353 – 1358.

12. R. Musunuri, , and J. A. Cobb, "Convergence of IBGP," in *Proc. of IEEE ICON Conference*, 2004.

13. M. G. Gouda, *Elements of Network Protocol Design*. John Wiley& Sons, 1998.

14. ——, "Protocol verification made simple: A tutorial," *Comput. Netw. ISDN Syst.*, vol. 25, no. 9, pp. 969–980, 1993.

15. R. Musunuri, , and J. A. Cobb, "Complete solution to IBGP stability," in *Proc. of IEEE ICC conference*, vol. 2, 2004, pp. 1177 – 1181.

16. T. G. Griffin, F. B. Shepherd, and G. Wilfong, "A safe path vector protocol," in *Proc. of INFOCOM conference*, 2000, pp. 490–499.

17. A. Arora and M. G. Gouda, "Distributed reset," *IEEE Trans. Comput.*, vol. 43, no. 9, pp. 1026–1038, 1994.

On the Sensitivity of Transit ASes to Internal Failures

Steve Uhlig*

Department of Computing Science and Engineering
Université catholique de Louvain, Louvain-la-neuve, B-1348, Belgium
suh@info.ucl.ac.be

Abstract. Network robustness is something all providers are striving for without being able to know all the aspects it encompasses. A key aspect of network design is the sensitivity of the network to internal failures. In this paper we present an open-source tool implementing the sensitivity model of [1], allowing network operators to study the sensitivity of their network to internal failures. We apply our methodology on the GEANT network, and we show that some of the routers and links of GEANT are sensitive to internal failures. Our results indicate that improvements can be made to the network design so as to reduce the risk of disruptions due to internal failures. Furthermore, we show great consistency between the results of the control plane and the data plane, indicating that applying the analysis on the control plane might be sufficient to provide insight into how to improve the resilience of the network to internal failures.

Keywords: network design, sensitivity analysis, control and data planes, BGP, IGP.

1 Introduction

Designing robust networks is a complex problem. Network design consists of multiple, sometimes contradictory objectives [2, 3]. Examples of desirable objectives during network design are minimizing the latency, dimensioning the links so as to accommodate the traffic demand without creating congestion, adding redundancy so that rerouting is possible in case of link or router failure and, finally, the network must be designed at the minimum cost. Recent papers have shown that large transit networks might be sensitive to internal failures. [4] has shown that a large ISP network might be sensitive to hot-potato disruptions. [5] extended the results of [4] by showing that a large tier-1 network can undergo significant traffic shifts due to changes in the routing. To measure the sensitivity of a network to hot-potato disruptions, [1] has proposed a set of metrics that capture the sensitivity of both the control and the data planes to internal failures inside a network.

To understand why internal failures are critical in a large transit AS, it is necessary to understand how routing in a large AS works. Routing in an Autonomous System (AS) today relies on two different routing protocols. Inside an AS, the intradomain routing protocol (OSPF [6] or ISIS [7]) computes the shortest-path between any pair of routers inside the AS. Between ASes, the interdomain routing protocol (BGP [8])

---

* This research was carried while the author was visiting Intel research Cambridge. Steve Uhlig is "chargé de recherches" with the FNRS (Fonds National de la Recherche Scientifique, Belgium)

T. Magedanz, E.R.M. Madeira, and P. Dini (Eds.): IPOM 2005, LNCS 3751, pp. 142–151, 2005.

is used to exchange reachability information. Based on both the BGP routes advertised by neighboring ASes and the internal shortest paths available to reach an exit point inside the network, BGP computes for each destination prefix the "best route" to reach this prefix. For this, BGP relies on a "decision process" [9] to choose its a single route called the "best route among several available ones. The "best route" can change for two reasons. Either the set of BGP routes available has changed, or the reachability of the next-hop of the route has changed due to a change in the IGP. In the first case, it is either because some routes were withdrawn by BGP itself, or that some BGP peering with a neighbor was lost by the router. In the second case, any change in the internal topology (links, nodes, weights) might trigger a change in the shortest path to reach the next hop of a BGP route. In this paper we consider only the changes that consist of the failure of a single node or link inside the AS, not routing changes related to the reachability of BGP prefixes.

In this paper, we propose an open-source tool allowing network operators to study the sensitivity of their network to internal failures. Contrary to [1] whose implementation of the sensitivity model is not available, our tool is freely available. We rely on the metrics proposed in [1] and extend the model by removing the limitations on the structure of the BGP sessions inside the AS as well as considering the complete BGP decision process [9]. Furthermore, while [1] studied the sensitivity of the control plane of a tier-1 AS, here we study the sensitivity of both the control and the data planes of the GEANT network.

Our study confirms that the metrics proposed in [1] provide insight into the sensitivity of the network to internal failures. More important is the necessity to confront the sensitivity analysis of the control and the data planes of [1], because the uneven traffic distribution towards destination prefixes [10, 11] might make the results of the control and data planes different. Our study of the GEANT network however indicates that the control and the data plane of this network have a similar sensitivity.

Note that for reasons of space limitation we do not describe in this paper the methodology used to build snapshots of the routing and traffic of an AS, but we refer the reader to [12].

The remainder of this paper is structured as follows. Section 2 introduces the building blocks of the sensitivity model. Section 3 presents the metrics to measure the control plane sensitivity. Section 4 applies these metrics to the control plane of GEANT. Section 5 then presents the metrics to measure the data plane sensitivity and Section 6 studies the sensitivity of the data plane of GEANT.

2   Network Sensitivity to Internal Failures

Let $G = (V, E, w)$ be a graph, V the set of its vertices, E the set of its edges, w the weights of its edges. A graph transformation δ is a function $\delta : (V, E, w) \rightarrow (V', E', w')$ that deletes vertex or edge from G. In this paper we consider only graph transformations δ that consist in removing a single vertex or edge from the graph. For consistency with [1], we denote the set of graph transformations of some class (router or link failures) by ΔG. The new graph obtained after applying the graph transformation δ on the graph G is denoted by $\delta(G)$. Due to space limitations, we restricted the set of graph transformations as well as the definition of a graph compared to [1], by not considering changes in the IGP cost. Changes to the IGP cost occur rarely in real networks, and never in the GEANT network. Our methodology however has no limitation on the set of graph transformations, IGP changes could be considered simply by extending our definition of a graph G with weights and adding the corresponding set of graph transformations.

To perform the sensitivity analysis to graph transformations, one must first find out for each router how graph transformations may impact the egress point it uses towards some destination prefix p. The set of considered prefixes is denoted by P. The BGP decision process $dp(v, p)$ is a function that takes as input the BGP routes known by router v to reach prefix p, and returns the egress point corresponding to the best BGP route. The *region index set RIS* of a vertex v records this egress point of the best route for each ingress router v and destination prefix p, given the state of the graph G: $RIS(G, v, p) = dp(v, p)$.

We introduced the state of the graph G in the *region index set* to capture the fact that changing the graph might change the best routes of the routers. The next step towards a sensitivity model is to compute for each graph transformation δ (link or router deletion), whether a router v will shift its egress point towards destination prefix p. For each graph transformation δ, we recompute the all pairs shortest path between all routers after having applied δ, and record for each router v whether it has changed its best BGP route towards prefix p. We denote the new graph after the graph transformation δ as $\delta(G)$. As BGP advertisements are made on a per-prefix basis, the best route for each (v, p) pair has to be recomputed for each graph transformation. It is the purpose of the *region shift function H* to record the changes in the egress point corresponding to the best BGP route of any (v, p) pair, after a graph transformation δ:

$$H(G, v, p, \delta) = \begin{cases} 1, \ if \ RIS(G, v, p) \neq RIS(\delta(G), v, p) \\ 0, \ otherwise \end{cases}$$

The *region shift function H* is the building block for the metrics that will capture the sensitivity of the network to the graph transformations.

To summarize how sensitive a router might be to a set of graph transformations, the *node sensitivity* η computes the average *region shift function* over all graph transformations of a given class (link or node failures), for each individual prefix p:

$$\eta(G, \Delta G, v, p) = \sum_{\delta \in \Delta G} H(G, v, p, \delta) \cdot Pr(\delta)$$

where $Pr(\delta)$ denotes the probability of the graph transformation δ. Note that we assume that all graph transformations within a class (router or link failures) are equally likely, i.e. $Pr(\delta) = \frac{1}{|\Delta G|}, \forall \delta \in \Delta G$, which is reasonable unless one provides a model for link and node failures. Further summarization can be done by averaging the *vertex sensitivity* over all vertices of the graph, for each class of graph transformation. This gives the *average vertex sensitivity* $\hat{\eta}$:

$$\hat{\eta}(G, \Delta G, p) = \frac{1}{|V|} \sum_{v \in V} \eta(G, \Delta G, v, p)$$

The *node sensitivity* is a router-centric concept that performs an average over all possible graph transformations. Another viewpoint is to look at each individual graph transformation δ and measure how it impacts all routers of the graph on average. The *impact of a graph transformation* θ is computed as the average over vertices of the *region shift function*:

$$\theta(G, p, \delta) = \frac{1}{|V|} \sum_{v \in V} H(G, v, p, \delta)$$

The *average impact* of a graph transformation $\hat{\theta}$ summarizes the information provided by the *impact* by averaging it over all graph transformations of a given class:

$$\hat{\theta}(G, \Delta G, p) = \sum_{\delta \in \Delta G} \theta(G, p, \delta) \cdot Pr(\delta)$$

3   Control Plane Sensitivity

[1] relied on worst-case and best-case sensitivities in their *region shift function*, to capture the uncertainty as to whether a graph transformation would lead to a change of the egress point of a route for sure or not, depending on the behavior of the actual tie-breaking rules of the BGP decision process. In this paper, the *region shift function* relies on the BGP decision process as it exists on most routers [9], corresponding to a situation in-between the worst-case and best-case ones used in [1]. All the metrics defined in this section will have *RM* in superscript to indicate that these metrics concern the *routing matrix*, i.e. the set of egress points that can be used to reach a destination prefix by each ingress router.

To capture the impact of a graph transformation on the number of prefixes that will have to change their egress point, we sum for each graph transformation, the values of the *region shift function* over all considered prefixes and divide it by the total number of prefixes:

$$H^{RM}(G, P, v, \delta) = \frac{1}{|P|} \sum_{p \in P} H(G, v, p, \delta)$$

This new function H^{RM} is called the *routing shift function* for the control plane.

Based on the *routing shift function* for the control plane, we can now define the routing sensitivity of routers to graph transformations: the *node routing sensitivity*. The *node routing sensitivity* η^{RM} is computed as, for each router, the sum of the values of the *routing shift function* (for the control plane) over all values of the graph transformations multiplied by the graph transformation probabilities:

$$\eta^{RM}(G, P, \Delta G, v) = \sum_{\delta \in \Delta G} H^{RM}(G, P, v, \delta) \cdot Pr(\delta)$$

Again, we consider that all graph transformations are equally likely so that $Pr(\delta) = \frac{1}{|\Delta G|}$. The *average node routing sensitivity* $\hat{\eta}^{RM}$ summarizes the node routing sensitivity by doing the average of the *node routing sensitivity* over all routers:

$$\hat{\eta}^{RM}(G, P, \Delta G) = \frac{1}{|V|} \sum_{v \in V} \eta^{RM}(G, P, \Delta G, v)$$

While the *node routing sensitivity* provides an average over all graph transformations, a desirable goal for network design is to try to minimize the impact of the routing shifts at any router. To know the worst graph transformation in terms of the routing shift at each node, we compute the *worst routing shift* η_{max}^{RM} for each node, i.e. the maximum of the *routing shift function* over all graph transformations:

$$\eta_{max}^{RM}(G, P, \Delta G, v) = \max_{\delta \in \Delta G} H^{RM}(G, P, v, \delta)$$

For network robustness, one does not only care about the impact of the graph transformations on any single router of the network, but also the impact of a specific node or router failure on the whole network. For this, the *routing impact of a graph transformation* θ^{RM} is computed as the average fraction of route shifts (H^{RM}) over all vertices:

$$\theta^{RM}(G, P, \delta) = \frac{1}{|V|} \sum_{v \in V} H^{RM}(G, P, v, \delta)$$

The *average routing impact* $\hat{\theta}^{RM}$ summarizes the *routing impact* by averaging its value over the set of graph transformations of each class:

$$\hat{\theta}^{RM}(G, P, \Delta G) = \sum_{\delta \in \Delta G} \theta^{RM}(G, P, \delta) \cdot Pr(\delta)$$

Network design is not only about trying to minimize the average impact of link and node failures, but also the impact of the worst failure inside the network. The *maximum routing impact of a graph transformation* θ_{max}^{RM} gives for each graph transformation, the largest value of H^{RM} over all possible vertices of the graph:

$$\theta_{max}^{RM}(G, P, \delta) = \max_{v \in V} H^{RM}(G, P, v, \delta)$$

4   Control Plane Sensitivity of the GEANT Network

In this section we apply the metrics defined in the previous section on the control plane of the GEANT network. For this study, we used the largest prefixes that account for 90% of the total traffic of GEANT during the 28 considered days, a total of 4911 prefixes.

Figure 1 presents the *routing impact of the graph transformations* (θ^{RM}) on the routers of the GEANT network. The left part of Figure 1 gives the impact of router failures while the right one gives the impact of link failures. Our study relies on 28 daily snapshots in the life of GEANT, so each error bar on the graphs of Figure 1 gives the min-average-max (indicated by a point, beginning of continuous line, end of continuous line) values over the 28 days of the study. For all figures that display on their x-axis either routers or graph transformations, the objects shown represented in the x-axis have been ordered by increasing values of their average impact or sensitivity over time. The y-axis of Figure 1 gives the *routing impact* in percentage of the considered prefixes that shift their egress point after the failure.

The left part of Figure 1 provides the *routing impact* of node failures. The average *routing impact* of node failures is very small, under 5%, for most of them. The worst

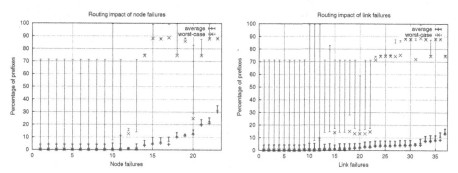

Fig. 1. Routing impact to graph transformations: router (left) and link (right) failures

node failure (θ_{max}^{RM}) impacts on average about 30% of the 4911 considered prefixes. To have a small average impact for a graph transformation means that the concerned routers or links are not used very often as egress points by the routers of the network. We can see that only 6 routers seem critical in the GEANT topology in that respect. In the GEANT network, some routers are mainly used to connect the NRENs (National and Regional research and Education Networks) to the network, not to provide connectivity outside the NRENs. These routers only attached to NRENs and not other peers are mainly ingress points and are not used much as egress points by other routers of the network. Their failure hence mostly impacts the connectivity with a few prefixes advertised by the concerned NREN. On the other hand, some routers can have a non-negligible routing impact in the network. The *worst-case routing impact* is a little more complex to understand than the average routing impact. The graph transformations having a small routing impact also have a small *worst-case routing impact* most of the time, except for one particular time bin (valid for router and link failures). The graph transformations that have the largest routing impact however have a large *worst-case routing impact* all the time, meaning that these graph transformations are critical for at least one router all the time. Improving the resilience of the network could hence be done by protecting these routers that might suffer from these highly disruptive graph transformations, or by splitting the best routes of these routers so reduce the impact of a single router or link failure.

Note that the observations made so far are highly related to the design of the GEANT network which relies a lot on hot-potato routing and where no BGP tweaking is made so as to split the set of best routes used to reach prefixes evenly among the available egress points of the network.

While the *routing impact* gives an average over the routers of the network, it is interesting to have a more detailed view at the individual sensitivity of each router of the topology to graph transformations. Figure 2 shows for each router its *node routing sensitivity* (η^{RM}), along with the *worst routing shift* (η_{max}^{RM}). Figure 2 shows that the average sensitivity is small, and more evenly balanced among the routers that the impact of the graph transformations on Figure 1. Only one router suffered from a large average *routing impact*, but only for a single time bin. So if we assume that all graph transformations are equally likely, the risk that a given router will suffer from big routing shifts is low on average. However, the *worst routing shift* (η_{max}^{RM}) give us another viewpoint. All

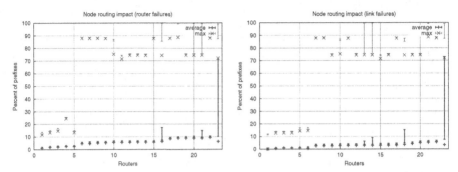

Fig. 2. Node sensitivity to graph transformations: router (left) and link (right) failures

except a few routers will suffer a very large routing shift (more than 70% of its routes) for at least one graph transformation, meaning that all the best routes of that router cross the concerned link or router. This implies that improvement in the design can be made by trying to spread the best routes over the available paths and egress points of the network to prevent a single link or router failure to have such a large impact on some routers.

Even though some graph transformations are more important than others (particularly router failures) when their impact is averaged over all routers, individual routers do not see wide differences in their average sensitivity to graph transformations. The situation for the *worst-case routing impact* (θ_{max}^{RM}) and the *worst-case node routing sensitivity* (η_{max}^{RM}) is quite different. Almost all routers on Figure 2 show a large *worst-case node routing sensitivity*, meaning that most routers are highly impacted by at least one graph transformation, even though on average each router is not much affected by graph transformations. This point to the fact that with BGP, large set of prefixes share the same egress point for a given ingress router. Hence it is highly likely that at least one router or link failure will affect an important egress point for any given router. Note that a few routers are not very sensitive to graph transformations. These nodes are actually those having external peerings, i.e. the routers most heavily used as egress points in the network. As these routers very often have as their best route one learned from an external peer, they are those less sensitive to disruptions that occur inside the network. The five routers that are the less sensitive to link and router failures are actually those that are most critical for all the rest of the network. This means there is room for improving the design of the network by reducing the criticality of these five routers, at least by splitting the best routes of the ingress routers more evenly between these five egress routers so that one failure does not impact so much some routers.

5   Data Plane Sensitivity

Let P be the set of destination prefixes and $I \in V$ be the set of ingress routers. The traffic demand M is an $|I| \times |P|$ matrix, whose elements $M(v, p)$ represent the amount of traffic that is received at ingress router v towards destination prefix p. The total inbound traffic received at an ingress router towards all destination prefixes of P is

$$T(v) = \sum_{p \in P} M(v, p)$$

In this paper we use one-day time bins for the traffic demand. We do not index all variables by the time to prevent unnecessarily cumbersome notations, but the reader must be aware that all variables are computed for each time bin. Similarly to the previous section, all metrics of this section have TM in superscript to indicate that they concern the traffic matrix.

As the *routing shift function* H^{RM} for the control plane, the *traffic shift function* H^{TM} gives for each prefix p, the amount of traffic entering ingress v that switches to other egress routers after a graph transformation δ. This is done by summing over all prefixes $p \in P$, the value of the *region shift function* H multiplied by the amount of traffic for the given (v, p) pair:

$$H^{TM}(G, P, v, \delta) = \frac{1}{T(v)} \sum_{p \in P} H(G, v, p, \delta) \cdot M(v, p)$$

The sensitivity of each ingress router to traffic shifts is represented by the *ingress node traffic sensitivity* η^{TM} and is computed as the sum over all graph transformations of the *traffic shift function* H^{TM} multiplied by the probability of the graph transformation δ:

$$\eta^{TM}(G, P, \Delta G, v) = \sum_{\delta \in \Delta G} H^{TM}(G, P, v, \delta) \cdot Pr(\delta)$$

Each transformation is again supposed to be equally likely. The *maximal ingress node traffic sensitivity* η^{TM}_{max} is, for each ingress node, the maximum of the *traffic shift function* over all possible graph transformations:

$$\eta^{TM}_{max}(G, P, \Delta G, v) = \max_{\delta \in \Delta G} H^{TM}(G, P, v, \delta)$$

Then the *average ingress node traffic sensitivity* $\hat{\eta}^{TM}$ gives the average of the *ingress node traffic sensitivity* computed over all ingresses, for each graph transformation:

$$\hat{\eta}^{TM}(G, P, \Delta G) = \frac{1}{|I|} \sum_{v \in I} \eta^{TM}(G, P, \Delta G, v)$$

The *traffic impact of a graph transformation* θ^{TM} measures the fraction of the traffic that shifts because of a graph transformation δ, averaged over all ingress points of the graph:

$$\theta^{TM}(G, P, \delta) = \frac{1}{|I|} \cdot \sum_{v \in I} H^{TM}(G, P, v, \delta)$$

θ^{TM} captures the change in the traffic matrix due to the graph transformation. The *maximal traffic impact* θ^{TM}_{max} of a graph transformation δ gives the maximum of the *traffic shift function* H^{TM} computed over the ingress nodes of the graph:

$$\theta^{TM}_{max}(G, P, \delta) = \max_{v \in V} H^{TM}(G, P, v, \delta)$$

The *average traffic impact* $\hat{\theta}^{TM}_{max}$ sums the *traffic impact of a graph transformation* θ^{TM} over all graph transformations δ multiplied by the probability of the graph transformation:

$$\hat{\theta}^{TM}(G, P, \Delta G) = \sum_{\delta \in \Delta G} \theta^{TM}(G, P, \delta) \cdot Pr(\delta)$$

6   Data Plane Sensitivity of the GEANT Network

In this section, we go to the data plane side of the sensitivity analysis. As traffic in
general seems to be unevenly distributed among the destination prefixes [10, 11], one
should not expect that the sensitivity analysis for the control plane be consistent with
the one of the data plane.

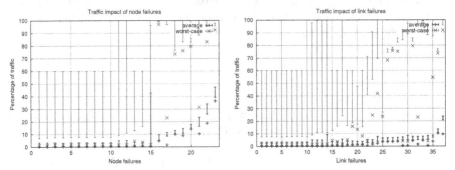

Fig. 3. Traffic impact to graph transformations: router (left) and link (right) failures

Figure 3 shows the *traffic impact of graph transformations*. As usual, the graph
transformations on the x-axis of Figure 3 have been ordered by increasing value of their
average *traffic impact* over the 28 daily snapshots. The impact of the graph transformations are similar for the data and the control planes. The average impact of the graph
transformations are small for most graph transformations. We can see that the most disruptive router failure has a slightly larger average traffic impact than its routing one,
about 39% of the traffic against 31.5% of the considered prefixes. But overall, the results for the traffic and routing impact are pretty much the same. The *worst-case traffic
impact* (θ_{max}^{TM}), as for the control plane, is smaller than 10% for 14 routers and 21 links
except for a single time interval. The consistency between the results for the control
plane and the data plane indicate that the distribution of the traffic among ingress-egress
pairs inside GEANT samples relatively well the distribution of the egress points found
by BGP. The traffic matrix does not seem to change much the routing sensitivity in
the GEANT network, at least for the largest 4911 prefixes capturing 90% of the traffic
during the 28 days we considered.

7   Conclusions

In this paper we proposed an implementation of the sensitivity model to internal failures
of [1]. Our version of the model is sensitive to any predicted change of the best BGP
route selected by a router, and does not rely on assumptions concerning the internal
BGP configuration of the network.

We applied the sensitivity analysis on GEANT to better understand its design and
robustness to internal failures. We showed that some of the routers and links of the
GEANT network are highly critical and sensitive to internal failures. This analysis has

implications on the protection that might be done inside the network to prevent critical router and link failures to create big disruptions in the network. Furthermore, we found consistency between the results of the control plane and the data plane, indicating that applying the analysis on the control plane might be sufficient to provide insight into the design of the network. We believe that large ISPs might benefit from carrying the same study as we did in this paper to improve their understanding of their network design choices.

Acknowledgments

We thank GEANT for making their routing and traffic data available. Thanks to Renata Teixeira for comments on earlier versions of this paper. This work was partially supported by the E-NEXT NoE funded by the European Commission.

References

1. R. Teixeira, T. Griffin, G. Voelker, and A. Shaikh, "Network sensitivity to hot potato disruptions," in *Proc. of ACM SIGCOMM*, August 2004.
2. R. S. Cahn, *Wide Area Network Design: Concepts and Tools for Optimisation*, Morgan Kaufmann, 1998.
3. W. D. Grover, *Mesh-Based Survivable Networks*, Prentice Hall PTR, 2004.
4. R. Teixeira, A. Shaikh, T. Griffin, and J. Rexford, "Dynamics of hot-potato routing in IP networks," in *Proc. of ACM SIGMETRICS*, June 2004.
5. R. Teixeira, N. Duffield, J. Rexford, and M. Roughan, "Traffic matrix reloaded: impact of routing changes," in *Proc. of PAM 2005*, March 2005.
6. J. Moy, *OSPF : anatomy of an Internet routing protocol*, Addison-Wesley, 1998.
7. D. Oran, "OSI IS-IS intra-domain routing protocol," Request for Comments 1142, Internet Engineering Task Force, Feb. 1990.
8. J. Stewart, *BGP4 : interdomain routing in the Internet*, Addison Wesley, 1999.
9. Cisco, "BGP best path selection algorithm," http://www.cisco.com/warp/public/459/25.shtml.
10. J. Rexford, J. Wang, Z. Xiao, and Y. Zhang, "BGP Routing Stability of Popular Destinations," in *Proc. of ACM SIGCOMM Internet Measurement Workshop*, November 2002.
11. S. Uhlig, V. Magnin, O. Bonaventure, C. Rapier, and L. Deri, "Implications of the Topological Properties of Internet Traffic on Traffic Engineering," in *Proc. of ACM SAC'04*, March 2004.
12. B. Quoitin and S. Uhlig, "Modeling the routing of an Autonomous System with C-BGP," *To appear in IEEE Network Magazine, special issue on interdomain routing*, 2005.

A NetConf Network Management Suite: ENSUITE

Vincent Cridlig, H. Abdelnur, J. Bourdellon, and Radu State

LORIA – INRIA Lorraine
615, rue du jardin botanique
54602 Villers-les-Nancy, France
{cridligv,festor,state}@loria.fr

Abstract. This paper presents a full NetConf-based network management suite. The described management suite allows to assess the potential of XML based management in complex configuration scenarios, validate NetConf operational characteristics, compare with existing models, verify the admitted ideas about XML and experiment the interoperability with CLI. We share here our implementation and deployment experience and discuss through illustrations the advantages and drawbacks of NetConf.

Keywords: XML-based network management, NetConf.

1 Introduction

XML-based network management is now a credible alternative to SNMP or CLI. Many efforts have been produced to move progressively from SNMP to XML ([6]) and ended up in the design of hybrid systems or complete XML-based management frameworks.

The reason why interest came to XML and its satellite technologies is that XML technologies offer many advantages (interoperability, existing tools, popularity) that we will illustrate along this paper. However it may also raise some unexpected practical drawbacks (CPU and bandwidth consumption).

This paper presents YencaP, a Python-based NetConf implementation. It also embeds some proposed extensions to the NetConf protocol and describes our approach for instrumenting a BGP software router (Quagga [1]).

The objectives that motivate our work are to provide an open-source NetConf agent, to test concretely a large scale XML-based network management solution, to test and practically validate the proposed methods of NetConf (such as subtree filtering) and also to provide a testbed for future proposed approaches.

This paper shares our experience in the use of a NetConf management suite for complex network management tasks. It first introduces the NetConf configuration protocol in section 2. Section 3 details the NetConf agent (YencaP) architecture, a YencaP module and the NetConf web-based manager. In section 4, we underline some related works. Section 5 concludes this paper and gives an overview of our future works.

T. Magedanz, E.R.M. Madeira, and P. Dini (Eds.): IPOM 2005, LNCS 3751, pp. 152–161, 2005.

2   Netconf

2.1   Protocol Overview

NetConf is an XML-based, configuration-oriented, network management protocol. It is an application level protocol and can rely on different existing protocol such as SOAP, BEEP or SSH. In NetConf, a manager queries an agent which runs on a remote device. Different kinds of queries called operations are available to pull/push XML-formated configuration data from/to the device. NetConf is designed to be scalable and to be light enough to be embedded in small devices.

A NetConf message belongs to a NetConf session. The latter is initiated when two peers, one acting as a manager and one as an agent, exchange their own capabilities in a *hello* message along with a session identifier from the agent. All following messages are requests and replies according to a RPC schema.

A NetConf request can contain different operations: get or get-config to retrieve configuration data, edit-config to modify data, copy-config to copy a full configuration, lock/unlock to gain unique access to information, close-session/ kill-session to shutdown a NetConf session. For more details on NetConf operations, please report to the draft [4].

NetConf provides the ability to define multiple configurations. For instance, *running* configuration is the configuration which is currently active on the device, while *candidate* configuration represents a configuration that may become active later and *Startup* is the configuration that must be loaded at boot time.

An important feature are the capabilities. A capability is a feature that a NetConf peer supports or not. NetConf defines a set of basic capabilities like *#candidate*, *#startup*, *#xpath*, *#writable-running* and authorizes the definition of new ones for flexibility. Capabilities are exchanged at the beginning of a new NetConf session.

Most NetConf advantages and drawbacks are inherent to XML technology [6] on which it relies. NetConf drawbacks relate to potential performance of XML parsing and XML data size compared to ASN.1 or CLI configuration.

3   NetConf Management Suite

3.1   NetConf Agent

Architecture. The design of our NetConf agent implementation, called YencaP, was led by the flexibility of the NetConf protocol design. It intends to follow these requirements: be modular, easy to use, fast to update, secure and efficient.

The YencaP architecture is organized in functional blocks. The low level blocks, except the lowest one, follow a layer-based architecture and basically relate to the XML nodes of a RPC request: network level, rpc level, NetConf operation level. *RPC layer* is responsible for (de)compression and en(de)cryption of NetConf requests. Figure 1 shows the lowest layer (SERVER) and the SLP (Service Location Protocol [5]) part. SLP allows YencaP agent to register as

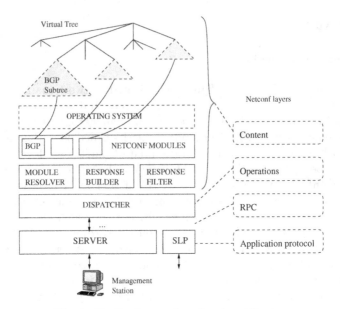

Fig. 1. NetConf agent functional architecture

a *Service Agent* to a *Directory Agent* so that a NetConf manager (SLP *User Agent*) can locate it.

Figure 1 also illustrates the upper functional blocks. *Dispatcher* delegates the request processing serially to the three upper layers. *Module resolver* is in charge of resolving what logical modules are able to treat the current request. It relies on the module registration process. When loaded, each module discloses the node it manages using XPath. For example, a BGP module may manage */netconf/routing/bgp* node. Based on that information, *Module resolver* can guess the set of concerned modules for any NetConf operations. By default, if no module is found, all modules are queried. *XML Response Builder* will query all modules from the module list to perform the request. Then, the local responses are checked and gathered to build a single global response. *XML Response Filter* filters the XML output in case of *get* or *get-config*. A module reply contains the whole module configuration. When the manager asked for a particular node set (via XPath or subtree filtering), the non matching node must be cut by *XML Response Filter*.

In order to allow agent extensibility, our experience leads to recommend the *command* design pattern that suits very well to implement the NetConf operations: *get-config*, *edit-config*, ... It allows to implement operation queues, undoable operations and to add new commands without changing the existing code. This is particularly helpful for vendors that may want to add new operations without any change to the NetConf stack. Combined with the *composite* design pattern, it is possible to easily implement macro-commands. For instance, the series (lock, edit-config, unlock) can be implemented as a *super-edit-config* macro-command, which would be a new NetConf operation and capability.

Extension features. YencaP also implements security services: authentication, encryption and access control. YencaP stores an access control policy based on RBAC model ([8]). Resources that are part of the permission definitions are described using XPath. When security is enabled on the agent, each NetConf request must be performed on behalf of a role (RBAC). That role is related to an encryption key. The key knowledge ensures the ownership of a manager to a role and a manager must be authenticated to get these keys. A manager must be allowed by the access control policy (RBAC user assignment relationships) to get one of them. Thus, *XML Response Filter* also filters the response of a *get/get-config* according to the role in use.

YencaP implements compression that can be optionally enforced. Some of our early tests show that the bandwidth gain can go up to 70%.

3.2   BGP Module Overview

We designed and implemented a complete module for the management of a BGP router (Quagga [1]) In our BGP Module implementation for the NetConf agent the most complex functionality was found in the bidirectional mapping from the configuration represented under the form of a XML document to a series of CLI commands. A Meta Language was designed to drive the CLI-XML translation. This language provides a full support for conditional expressions, loops and dynamic types. For illustrative purposes, we will show a subpart of the command "ip prefix-list" allowing us to introduce an example showing how this translation works. The command syntax will be:

```
ip prefix-list NAME (seq NUMBER)? (permit|deny)
```

The *Meta Language* (in Figure 2) consists of two different node types 1) the ones to specify instructions (i.e. `<parser>`, `<command>`, `<children>`, etc) and 2) the ones standing for the desired XML document (i.e. the main node and all the children of a `<children>` node).

In order to generate the **CLI-to-XML translation**, first the *CLI command* is parsed to create a structure of primitives types, which is shown in *CLI-to-Primitives Types parsing* at Figure 2. Next the *Meta Language* is used to generate the XML Node document using the `<parser>` nodes on it, where the main objective is to keep these two structures aligned. For example, specifying that the type ("valuetype" attribute) of the `<prefix-list>` is a list and for which, each item in this list will re-print the `<prefix-list>` node. This is done by exploring the whole *Meta Language* document trying to keep that alignment and in case this is possible, it will create the corresponding nodes.

To generate the **XML-to-CLI translation**, the `<command>` and `<followcommand>` from the *Meta Language* and the *XML command* are used (Figure 2). Since each different scope of the *XML Command* document represents a different context of the CLI configuration or even a part of a CLI command, the above nodes represent different needs i) the `<followcommand>` stands for creating subpart of the CLI command, the one shown in the *Meta Language* at the `<prefix-list>` will be **ip prefix-list 100** which should be later appended to the

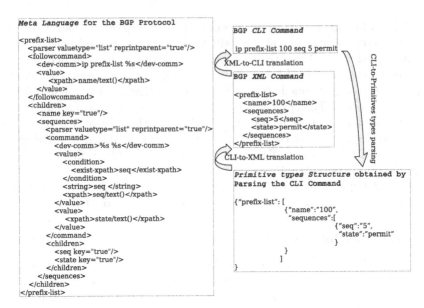

Fig. 2. CLI-XML Mapping

commands generated by its children and ii) the <command> stands for creating a complete command, which for the case of the <sequences> node will generate the CLI command **seq 5 permit** and will concatenate the followcommand already generated to it.

To illustrate the proceeding of the evaluation of a *command* node, consider the <command> node at the *Meta Language* in the figure 2. The text in the <dev-comm> will be the command generated after formatting it with the values acquired from the <value> nodes. In this case, one of the <value> contain a <condition> which will be evaluated in the *XML Command* at the current scope where this node was specified. If this is validate to true, a string containing the "seq" and the value of the <seq> will be created and that will replace the first "%s" of the <dev-comm> and the second will be the value extracted from the <state> node as specified in the next <value>.

3.3   NetConf Manager Overview

Manager architecture. The web-based NetConf management application, illustrated on Figure 3, is divided into two main layers, the *HTTP layer* and the *NetConf layer*.

The *HTTP layer* is composed of a *HTTP server* and a *session manager*. The *HTTP server* is able to process incoming HTTP requests from remote users, serving both static and dynamic content. In the case of a NetConf request, it forwards all needed information to the *Session Manager* and then sends back the response to the remote user. The *Session Manager* is in charge of several HTTP sessions. Each remote manager is bound to a HTTP session. When a user request is received, the HTTP server dispatches the information to the appropriate HTTP session and retrieves the response for this request.

The *NetConf layer* is made of two functional blocks: the set of HTTP sessions and the set of NetConf sessions. A *HTTP Session* is created for every remote user and can manage one or more NetConf session. However, a NetConf session belongs to one and only one HTTP session. An incoming request is forwarded to the specific NetConf session implemented as a thread. The NetConf client queries the related NetConf agent and retrieves the NetConf response.

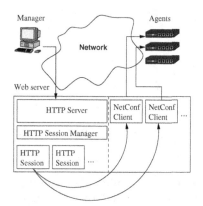

Fig. 3. Manager functional architecture

Manager web interface. Figure 4 illustrates the web-based NetConf manager interface. It allows administrators to send NetConf requests without any a priori knowledge about the NetConf message format. Based on the information collected through a HTML form, the management application automatically generates a well-formed NetConf request, sends it to the specified agent and then returns the response. Remote users only need to know the agent to be managed and the query to be performed.

3.4   Subtree Filtering Versus XPath

The *get-config* operation [4] allows two different node selection approaches. Subtree filtering is the default one and is NetConf specific. XPath capability is the second approach and makes use of regular XPath expressions to retrieve configuration nodes. It is natural to assess the performance of theses selection methods since a common hypothesis is that subtree filtering might be more performant.

In an implementation point of view, each filtering method has its own benefits and drawbacks. First, XPath capability is really simple to implement at first sight. Applying an XPath request on a DOM document is a simple operation. However, when the XPath expression is complex (not absolute for instance), it starts to become more complicated to find the right part of the data model. It requires to build the whole DOM document before applying the request. A

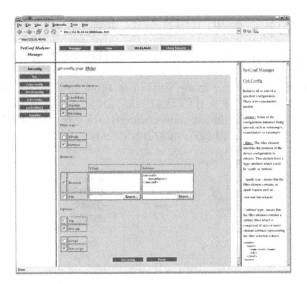

Fig. 4. Manager Web Interface

relative XPath expression doesn't allow a dichotomic algorithm since the root XML tags are not known. When the XPath expression is simple, it is easy to build a sub part of the DOM tree and apply the XPath expression on it. The use of hash tables does not help because some parts of the global data model may have the same tag names. Therefore, it is not always easy to bind a XPath expression to a subtree. The use of namespaces can help a lot to build the response more efficiently since it uniquely identifies the concerned subtrees of the global XML tree. Second, even if subtree filtering requires to be implemented from scratch, it makes it easier to efficiently retrieve the data compared with relative XPath expressions. For absolute XPath, the algorithm complexity is totally equivalent. However, contrary to XPath, subtree filtering requires an huge effort to learn its rules in details even for experienced XML users.

In order to compare execution time of both methods, we prepared two couples of equivalent requests. The first couple is illustrated in Figure 5. In the subtree example, since `<login>bob</login>` is a content match node, all users whose login equals 'bob' will be selected. Also, `<permissions/>` is a selection node, so all permissions will be selected in the output. In the XPath example, a node set will be selected and will be merged in a single document output. Note that in this example, whereas the weight of the rpc with subtree filtering is 377 bytes, the weight with the XPath method is 288 bytes. Hence a rate of 76% between the size of the two requests. The second couple get-config selects */netconf/security/rbac/permissions/permission/scope*. The rate for the second example is 68% favorable to XPath. These are arbitrary but realistic example.

The results of our experiments are the following. With the first get-config couple, the average processing time on 80 requests is 1.58 ms for subtree filtering and 1.44 ms for XPath. With the second get-config couple, the average processing

```
<filter type="subtree">
  <netconf>
    <security>
      <rbac>
        <users>
          <user>
            <login>bob</login>
          </user>
        </users>
        <permissions/>
      </rbac>
    </security>
  </netconf>
</filter>

<filter type="xpath">
  /netconf/security/rbac/users/user[login='bob']
  | /netconf/security/rbac/permissions
</filter>
```

Fig. 5. Equivalent (Subtree filtering/XPath) get-config requests

time on 80 requests is 1.14 ms for subtree filtering and 1.29 ms for XPath. In our implementation, subtree filtering takes longer when different nodes must be merged.

However, the filtering time is comparable for both methods. On the one hand, the advantage for subtree filtering is that it makes it easier than XPath to produce a complex customized output document and helps to define a visual template of the desired output. One the other hand, XPath is easy to use, less verbose, largely well-known and having a strong support from the XML community. Another advantage is the large supporting libraries for XPath.

4   Related Works

The interest for web-based network management has increased for many reasons over the past ten years. First, a large panel of available tools appeared around the XML technology, allowing fast document processing (transformation with XSLT), parsing (SAX, DOM) and modeling (XML Schema). Some XML protocols also emerged, for instance SOAP, and provide a high level of interoperability among web services. Motivations and experiences of the use of XML technologies for network management is proposed in [10].

In this context, [7, 11–13] proposed a set of approaches relying on XML and SNMP. There are basically two models: native and hybrid models. While native model means pure XML or pure SNMP communications, hybrid models allow first an SNMP manager to manage an *XML-Based Management* (XBM) agent and second, an XBM manager to manage an SNMP agent.

A pure XML approach was proposed with Junoscript [9], which is very close to NetConf [2]. However, while NetConf proposes to use generic operations, Junoscript operations are data dependant.

In [3], we proposed an integrated security framework for NetConf. That framework is based on Role-Based Access Control, where requests are issued on behalf of a role, not of an user.

5   Conclusion

This paper describes a first open-source implementation of a NetConf management suite with extended features. Its modular architecture encourages developers to extend its supported services (BGP module, LOG module, ...). It supports encryption, authentication and access control as part of a global security architecture for NetConf. Also compression is supported to optimize bandwidth consumption.

This implementation is one more proof of the potential of XML technologies in network management systems. We experimented fast but efficient development process with XML and Python and benefited of existing XML technologies and supporting open-source Python modules. Furthermore, portions of code related to XML technologies (XSL, XML configuration documents, XPath, XML-Encryption) used in YencaP are independent of the programming language (Python). Thus a significant part of the code is outsourced in XSL. This proves interoperability and led us to some future approaches like *XSL pushing* which is really close to mobile code approaches.

We plan to continue our experiments on the framework: compression, encryption, access control, subtree filtering versus XPath. These criteria will impact the CPU and bandwidth consumption, response time and ability to support overload. Some early tests shown us that the agent can treat a figure of the order of 3 requests per second. Also compression can decrease the size of a XML document up to 70% which is considerable if a copy-config must be sent to millions of managed home gateways or mobile phones.

The YencaP management suite can be downloaded from our website: *madynes.loria.fr*.

References

1. Quagga Routing Suite. http://www.quagga.net.
2. M.-J. Choi, H.-M. Choi, J.W. Hong, and H.-T. Ju. XML-Based Configuration Management for IP Network Devices. *IEEE Communications Magazine*, 42(7):84–91, July 2004.
3. V. Cridlig, R. State, and O. Festor. An Integrated Security Framework for XML based Management. In *Proceedings of the Ninth IFIP/IEEE International Symposium on Integrated Network Management (IM 2005)*, IFIP Conference Proceedings, May 2005.
4. R. Enns. NETCONF Configuration Protocol. Internet Draft, work in progress, June 2005.

5. E. Guttman, C. Perkins, J. Veizades, and M. Day. Service Location Protocol, Version 2. STD 62, http://www.ietf.org/rfc/rfc2608.txt, June 1999.
6. S. Holzner. *Inside XML*. Pearson Education; 1st edition, November 2000.
7. T. Klie and F. Strauß. Integrating SNMP Agents with XML-Based Management Systems. *IEEE Communications Magazine*, 42(7):76–83, July 2004.
8. R. Kuhn. Role Based Access Control. NIST Standard Draft, April 2003.
9. T. Mauro. *JUNOScript API Guide for JUNOS Release 6.1.* Juniper Networks, 1194 North Mathilda Avenue. Sunnyvale, CA 94089, USA, sonia saruba edition, September 2003.
10. L.E. Menten. Experiences in the Application of XML for Device Management. *IEEE Communications Magazine*, 42(7):92–100, July 2004.
11. Y.-J. Oh, H.-T. Ju, M.-J. Choi, and J. W. Hong. Interaction Translation Methods for XML/SNMP Gateway. In Metin Feridun, Peter G. Kropf, and Gilbert Babin, editors, *Proceedings of the 13th IFIP/IEEE International Workshop on Distributed Systems: Operations and Management, DSOM 2002*, volume 2506 of *LNCS*, pages 54–65. Springer, October 2002.
12. F. Strauß and T. Klie. Towards XML Oriented Internet Management. In Germán S. Goldszmidt and Jürgen Schönwälder, editors, *Proceedings of the Eighth IFIP/IEEE International Symposium on Integrated Network Management (IM 2003)*, volume 246 of *IFIP Conference Proceedings*, pages 505–518. Kluwer, March 2003.
13. J.-H. Yoon, H.-T. Ju, and J. W. Hong. Development of SNMP-XML Translator and Gateway for XML-based Integrated Network Management. *International Journal of Network Management*, 13(4):259–276, 2003.

Rtanaly:
A System to Detect and Measure IGP Routing Changes

Shu Zhang[1] and Katsushi Kobayashi[1]

National Institute of Information and Communications Technology, Japan

Abstract. Routing changes of the interior gateway protocol (IGP), especially unexpected ones, can significantly affect the connectivity of a network. Although such changes can occur quite frequently in a network, most operators have hardly noticed them because of a lack of effective tools. In this paper, we introduce Rtanaly, a system to (i) detect IGP routing changes in real-time and instantly alert operators of the detected changes, (ii) quantify routing changes over the long term to provide operators with a general view on the routing stability of a network, (iii) estimate the impact of routing changes, and (iv) help operators troubleshoot in response to unexpected changes. Rtanaly has the following features: (i) it supports all three widely deployed IGPs - OSPFv2, OSPFv3, and IS-IS, (ii) it uses a completely passive approach, (iii) it visually displays the measurement results, and (iv) it is accessible through the web. We present the results of measurements that we have performed with Rtanaly as well as some observed pathological behavior to show its effectiveness. We have released the first version of Rtanaly as free software and its distribution is based on a BSD-style license.

1 Introduction

The Internet plays an increasingly important role in our society. People not only frequently obtain information that they use in their daily lives through the Internet, but also use it for critical tasks such as investment, banking, and shopping transactions, which require higher levels of reliability. However, the infrastructure of the Internet is still fragile and reachability can be seriously degraded by small failures. In this paper, we focus on interior gateway protocol (IGP) routing changes which can directly affect the performance of an IP service network.

We use the term "IGP routing changes" to refer to changes that can be observed in the link state update messages of an IGP in an operational network. In a routing domain, all routers calculate intra-domain routes based on link state update messages, so changes in these messages directly affect the connectivity of a network. IGP routing changes can be divided into two categories. The first includes changes due to network maintenance and traditionally these changes are considered unavoidable[1]. The second type of IGP routing change includes those arising from unknown causes, most of which are network failures. In our work, we mainly deal with such unexpected changes.

The results of measurements we have performed and of other studies [1] [2] [3] [4] have shown that although network operators are rarely aware of them, unexpected IGP

---

[1] Though mechanisms such as graceful restarting to minimize the influence of such changes are being developed

T. Magedanz, E.R.M. Madeira, and P. Dini (Eds.): IPOM 2005, LNCS 3751, pp. 162–172, 2005.

routing changes can occur in a service network, and sometimes they can occur quite frequently. These changes are difficult to detect because they tend to occur intermittently. In many cases, by the time an operator gets a report that a routing problem has occurred and starts troubleshooting, the problem has disappeared. Thus, an effective system to help operators monitor such anomalies is crucial to provide high quality connectivity service.

In this paper, we present Rtanaly (RouTe ANALYsis), an open system we developed to detect and measure IGP routing changes. The system (i) detects IGP routing changes in real-time and alerts operators of the changes, (ii) gives the operator a general view on the routing stability of a network by quantifying routing changes over the long term, (iii) estimates the impact of IGP routing changes on the network, and (iv) helps operators troubleshoot problems. Furthermore, by making this system open software, we expect to make it easier for other parties in the research community to perform more measurements on IGP routing changes in different networks, which will allow potential IGP routing problems to be identified more quickly and easily. While we assume Rtanaly will be used in large or relatively large service networks, it can also be deployed in small-scale networks, such as a campus network, to monitor the reachability of the network. Rtanaly is free software and its distribution is based on a BSD-style license. To the best of our knowledge, it is the only open source software at this point that deals with IGP routing changes.

Commercial and academic efforts have gone into developing an IGP monitoring system. Route Explorer and RouteDynamics are commercial products released by Packet Design and Ipsum Networks. However, little is known about the specifics of these products. In [5], Baccelli et al. presented the design and the implementation of an OSPFv2 [6] monitoring system, but did not focus on routing changes. In addition, the implementation was neither tested in an operational network nor made public. In [7], Shaikh et al. described the design of another OSPFv2 monitoring system. While the architecture of this work is somewhat similar to that of our system, we believe Rtanaly offers several advantages:

- It has been released as open source software, which makes it possible for more extensive measurements to be performed in other networks.
- The impact of observed IGP routing changes can be estimated.
- Results can be clearly visualized, and are accessible through a web browser.
- It provides more support for troubleshooting.
- OSPFv3 [8] and IS-IS [9], as well as OSPFv2, are supported.

The rest of this paper is organized as follows. In Section 2, we describe the methodology used in Rtanaly. Section 3 presents the architecture and functions of Rtanaly. Section 4 shows the results of measurements done with Rtanaly. We conclude in Section 5.

2 Methodology

2.1 Link-State Routing Protocol

The link-state routing protocol is preferred by most ISPs because of its flexibility, robustness, and efficiency. In a typical link-state routing protocol, the link state update message is flooded throughout a network to disseminate routing information. All routers located in the same routing domain calculate their own routing table based on this information. When there is any change in the network topology, the advertising router will generate a new update message and reflood it. By comparing a new update message with the previous one, we can figure out what kind of changes have occurred.

2.2   OSPF

In OSPF, link state advertisements (LSAs) are used to disseminate routing information. Currently, five kinds of LSA are used most often in OSPFv2. They are: Router-LSA, Network-LSA, Network-Summary-LSA, ASBR-Summary-LSA and AS-external-LSA.

OSPFv3 was developed for IPv6 routing. In OSPFv3, two new LSAs are defined in addition to those in OSPFv2[2]: Link-LSA and Intra-Area-Prefix-LSA.

2.3   IS-IS

In IS-IS, the link state protocol data unit (LSP) is used to flood the routing information. An LSP consists of various triples of type, length, and value (TLV). By analyzing TLVs that are related to the route calculation, we can figure out what kind of routing changes have occurred in IS-IS. Note that not all TLVs affect the network reachability. The TLVs which need to be examined are the IS neighbors TLV, IP interface address TLV, IP internal reachability TLV, IPv6 reachability address TLV, extended IS reachability TLV, and extended IP reachability TLV.

2.4   Estimation of Impact on Network Reachability

When a routing change occurs, the operators will want to know (i) whether the network reachability will be affected by the change, and (ii) how the reachability will be affected if it is. Rtanaly is designed to answer these questions, and it estimates the impact of a routing change by analyzing the intra-domain path (IDP).

We define the IDP as the path a packet traverses from one router to another towards a destination in a routing domain. When any routing change occurs, the shortest-path tree (SPT) of the network for a specific router may or may not change. By comparing the SPTs before and after a routing change, we can figure out whether the SPT has changed, and if it has, how the IDPs from a specific router to other routers have changed. The IDP changes fall into the following categories:

1. The path changes. Its cost may, or may not change.
2. The path does not change, but the cost changes. Usually this derives from a change in the cost of a link.
3. The path becomes unreachable from a reachable state (a router disappears from the SPT).
4. The path becomes reachable from an unreachable state (a router reappears in the SPT).

---

[2] In OSPFv3, different names are used for the Summary-LSAs (Inter-Area-Prefix-LSA and Inter-Area-Router-LSA), but the functions are the same

3 Rtanaly

3.1 Architecture

Rtanaly consists of three components: RA-slaves, the RA-master, and the RA-webstat. As shown in Figure 1, each RA-slave is an IGP data collector that collects raw routing messages and transfers them to the RA-master. The RA-master analyzes the data received from the RA-slaves and generates an alert when it detects an anomaly. The RA-webstat functions on the same node as the RA-master and is the component that operators use to view measurement results regarding observed routing changes through a web browser.

In addition, Rtanaly utilizes the reliable data transmission protocol (RDTP), which we designed to ensure reliable transfer of routing messages from the RA-slaves to the RA-master.

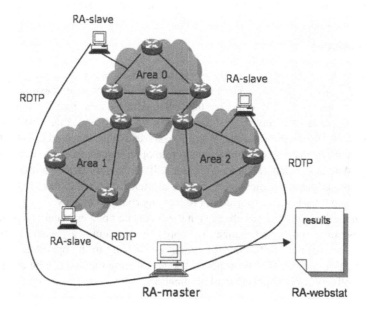

Fig. 1. Rtanaly architecture

3.2 RA-Slaves

The RA-slaves collect IGP routing messages at different points of the network and transfer the collected data to the RA-master. To minimize the possibility of inflicting any damage on the monitored network, we use a passive approach to collect the link state update information. As a result, each RA-slave must be connected to a place where the link state update information can be received, such as an Ethernet segment which is configured as a part of the routing domain, or use port mirroring technology[3]. Note

---

[3] Although this approach imposes some limitations on the deployment of an RA-slave, it can be easily applied in most networks

that with this approach, the complete link state database cannot be built up after Rt-analy is started until the refresh messages of all LSAs or LSPs are sent, which can take tens of minutes. However, because this delay occurs only at the very beginning of the measurement phase, we believe the affect is not significant[4].

An RA-slave uses the libpcap library[5], which is widely used to collect packets over a shared link such as an Ethernet link or a mirrored port, to capture the routing messages. When an RA-slave is started, it begins data collection and uses RDTP to transfer the collected messages to the RA-master. Sometimes an RA-slave cannot send messages to the RA-master because of a network failure, such as a routing problem or link failure. In this case, the RA-slave will hold the unsent routing messages and restart transferring them when the connection is restored.

In some cases, an RA-slave can be cut off from the remaining part of the monitored network and some update messages can be lost. To obtain a complete view of the network status, it is desirable to deploy more than one RA-slave in different parts of the network (e.g., in different subnets of an area or in different areas).

3.3   RA-Master

The RA-master is the protocol analysis engine. It receives routing messages from RA-slaves and analyzes the messages to find if any routing changes have occurred. The RA-master uses the RDTP to receive the collected messages from the RA-slave(s). It should be capable of receiving and analyzing data from more than one RA-slave. If data transfer between the RA-master and RA-slaves is disconnected unexpectedly, the RA-master starts a new session and waits for an open request from the RA-slaves.

The RA-master can detect a new LSA or LSP. It can also detect an expired LSA or LSP. When analyzing the messages, the RA-master discards duplicated messages and the refresh updates because they do not reflect any routing changes.

When the RA-master finds changes in the link state updates, it can: (i) instantly notify the operators of detected changes, (ii) notify the operators only when changes occur frequently, or (iii) not notify the operators at all. It is up to the operators to decide in which situation they want to be notified. The notification method can be email, syslog, etc. Currently, we only support email notification.

3.4   RA-Webstat

The RA-webstat provides a web interface that lets operators view the measurement results, investigate a specific flapping LSA or LSP, and so on. Specifically, it shows the following information.

Statistical results throughout the measurement period. These include statistics regarding all OSPF LSAs or IS-IS LSPs. In this way, operators can obtain a general view of the routing changes occurring in a network.

---

[4] It is possible to obtain the whole link state database from a neighbor through SNMP when Rtanaly is started, but we did not implement this method

[5] Packet capture library, see http://www.tcpdump.org

Statistical results for each day. These are the daily statistics for OSPF LSAs or IS-IS LSPs which enable operators to know what is happening on a specific day. Real-time analysis results are also provided. In addition, a ranking of LSAs or LSPs sorted by the number of changes is shown.

Statistical results for a specific LSA or LSP. If operators find any unexpected change of an LSA or LSP, they can investigate it in more detail. Both graphical statistics for the specific LSA or LSP and textual results on how it has changed are shown. With these results, operators can figure out which link or router caused a problem and start further troubleshooting based on this information.

Impact estimation. Whether any IDPs from a specific router to other routers has changed will be shown. If an IDP has changed, how it has changed will be displayed.

3.5   RDTP

The RDTP is used by RA-slaves and the RA-master to transfer collected routing messages. We took the following considerations into account in its design.

- It must be able to transfer the data collected at each RA-slave to the RA-master in real-time.
- It must ensure that the data sent to the RA-master is completely received. This requires explicit acknowledgments from the RA-master.

If the data cannot be transferred from an RA-slave to the RA-master because of any network failure, the RA-slave must hold the data (even if its connection with the RA-master times out) and periodically send open requests to the RA-master until the connection is restored. When the connection resumes, the RA-slave resends the holding data to the RA-master before sending any newly collected data. This mechanism guarantees that all collected data will be reliably transferred to the RA-master in spite of any network failure and requires sequencing of the data.

The RDTP uses TCP as its transport protocol. As shown in Fig. 2, the message sent from an RA-slave to the RA-master is composed of one byte for the version number (currently 1), one byte for the data type (1 for the libpcap format data), four bytes for the sequence number, and four bytes for the data length followed by the collected data. Further discussion of this protocol is beyond the scope of this paper[6].

3.6   Implementation and Availability

As we mentioned in Section 3.2, we use the libpcap library to capture the routing messages. The libpcap library also provides a function which can be used to do offline data analysis. We use RRDtool [10] to generate the statistical graphs. The RA-slave and RA-master components are implemented in the C language. The RA-webstat is written in

---

[6] In fact, RDTP is not directly related to the analysis of IGP routing changes. It can also be used in other measurements which collect data in real-time from multiple locations and gather the data in one place

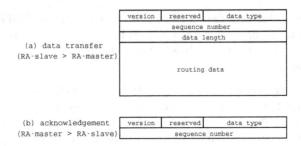

(a) data transfer
(RA-slave > RA-master)

(b) acknowledgement
(RA-master > RA-slave)

Fig. 2. RDTP message format

Perl script. Version 0.1 of Rtanaly was released in 2004 [11]. Although we have implemented most of the functions described in this paper, the support of RDTP is still a work in progress. The behavior of Rtanaly has been confirmed on Linux and BSD variants such as FreeBSD. Rtanaly should also work on other UNIX platforms because we provide a "configure" script generated by the GNU autoconf.

4   Measurements Using Rtanaly

4.1   General Results

Here we show results from our measurements done for the WIDE Internet [12], a national academic network in Japan. The measurement began in August 2000 and has lasted for more than four years. Figure 3 is a graph generated by the RA-webstat. It shows the number of changes observed in all intra-domain LSAs (type 1-4) each day during the period. This gives us a general view of the routing changes occurring in the WIDE Internet. We can see that IGP routing changes occur routinely on the WIDE Internet. Although the number of changes has been relatively low for most of the period, there have been days when the number spiked to very high levels, sometimes reaching 11,000 per day. Here we need to point out that some of these changes were due to normal network maintenance, but still the number is much higher than what we had expected.

Figure 4 shows a one-day example of LSA changes observed in the Router-LSA of a router located in Tokyo. Table 1 shows part of the detailed analysis generated by

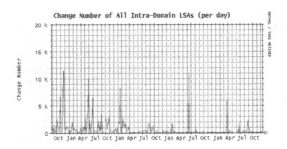

Fig. 3. Number of changes observed in all intra-domain LSAs

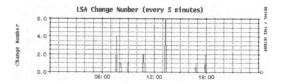

Fig. 4. Number of changes observed in a Router-LSA in one day

Table 1. Detailed analysis of a Router-LSA

Time	Seq. No.	Link ID	Link Data	Type	Cost
13:19:07	8000482c	-203.178.142.128	255.255.255.224	3	1
13:19:12	8000482d	+203.178.142.128	255.255.255.224	3	1
13:19:17	8000482e	-203.178.142.128	255.255.255.224	3	1
13:19:23	8000482f	+203.178.142.128	255.255.255.224	3	1
13:19:29	80004830	-203.178.142.128	255.255.255.224	3	1

the RA-webstat which indicates how this LSA changed. In this table, '+' indicates the addition of a link compared with the last instance of the LSA and '-' indicates the opposite. The type of 3 means the link is a stub one. Through this analysis we know that one of the router's interfaces repeatedly changed its state between up and down.

Figure 5 shows the number of daily LSP changes observed during a 9-month period in Abilene [13], which is the backbone network of Internet2 and uses IS-IS as the IGP. Abilene is a production network and consists only of 11 routers, but still the number of changes was quite high from time to time, with the maximum of 4299 times per day. Further study has shown that most of the changes were caused by link problems.

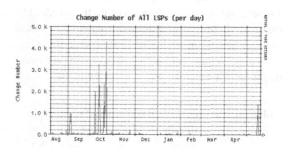

Fig. 5. Number of LSP changes observed in Abilene

4.2 Pathological Changes

Here we present some pathological changes that we observed during the measurements. We categorize these changes by their frequency and persistency.

Relatively Frequent Short-Term Oscillation. Figure 6 is an example of the most typical routing changes which occurred relatively frequently over a short term. We can see that during one day in May 2004, this Router-LSA changed twice in the early morning, declaring the down and up of two links in 10 seconds. It began the up/down advertisements again from noon and continued oscillating for about 4 hours. In our investigation,

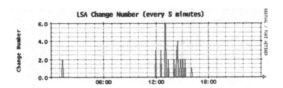

Fig. 6. Relatively frequent short-term oscillation

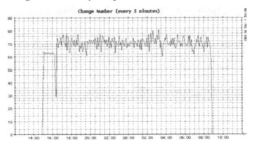

Fig. 7. Frequent short-term oscillation

of all different oscillation patterns, changes like this were observed most often. Usually they were caused by network congestion, but they could also be due to layer 2 problems.

Frequent Short-Term Oscillation. Figure 7 is an example of a serious oscillation. It shows the number of LSA changes for an L3 switch within 24 hours in 2004. As we can see, this router-LSA kept advertising the up/down of two links at a rate of about 70 times every 5 minutes, and in total it changed more than 16,000 times in 18 hours. We observed this kind of oscillation several times during our measurements, and most cases were caused by misconfiguration where the same router-ID was used on two different routers. Changes leading to this oscillation pattern can also be caused by p2p interface or link problems, and in this case, we will see two changing LSAs.

Long-Term Oscillation. Figures 8 and 9 show one-day examples of relatively frequent and less frequent long-term oscillation. Figure 8 is the result for a router located in San Francisco. We can see that the router originated changing LSAs relatively frequently, and the oscillation lasted for more than five months. Figure 9 is for a router located in Kyoto. We can see that the changes were relatively infrequent, with only several times a day, but the oscillation persisted for about two months. Oscillations like these were usually caused by interface or link problems.

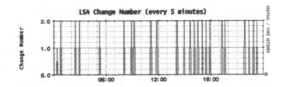

Fig. 8. Relatively frequent long-term oscillation

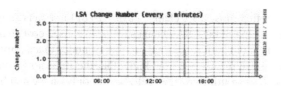

Fig. 9. Less frequent long-term oscillation

Although we observed many other interesting events during our measurements, we do not show all of them here due to space limit. More statistical results from Rtanaly are available in [11].

4.3 Impact Estimation

As an example of impact estimation, we analyzed the OSPF data collected from the backbone area of the WIDE Internet on December 20th, 2004. As shown in Fig. 10, a lot of OSPFv2 LSA changes occurred that day. The root we used in the SPT calculation was a router near the location where we collected the data.

First, we analyzed the node number change in the SPTs before and after an LSA change. We found that most (85.7%) of the LSA changes did not cause any change in the number of nodes. Fifteen (10.2%) changes caused fewer than five nodes to become reachable or unreachable, and only six of the total 147 changes caused more than five nodes to become reachable or unreachable.

We also analyzed how the SPT was affected by each routing change. We show the number of cost and nexthop changes caused by each LSA change in Fig. 11. From this

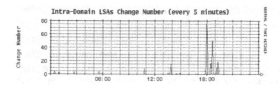

Fig. 10. Number of Type 1-4 LSA changes on December 20th, 2004

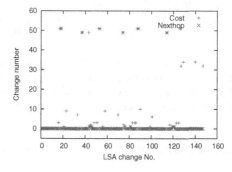

Fig. 11. SPT changes

graph, we can see that most LSA changes (70.7%) did not cause any cost or nexthop change. However, some of the LSA changes (10.2%) caused cost changes to more than 27 nodes, which is about half of the total number of nodes in a complete SPT.

Through the above results, we now know that the impact of routing changes also depends heavily on the routing change itself, but not only on the number of the changes.

5   Conclusion

Although previous work has shown that unexpected IGP routing changes can occur in a network, and sometimes occur quite frequently, network operators have hardly noticed them due to a lack of an effective monitoring and measuring system. In this paper, we have described Rtanaly, which not only detects and alerts operators of IGP routing changes, but also measures the overall routing changes that occur in a network over a long term and estimates the impact of routing changes. Rtanaly uses a passive approach to monitor the link state update messages of IGPs and supports all three widely deployed IGPs. Our long-term measurements on different networks have proven the effectiveness of Rtanaly.

We have released the first version of Rtanaly and plan to release the next one in 2005. RDTP will be implemented in the forthcoming version.

References

1. A. Shaikh, C. Isett, A. Greenberg, M. Roughan, and J. Gottlieb, "A case study of ospf behavior in a large enterprise network," in *Proceedings of ACM Internet Measurement Workshop*, 2002. [Online]. Available: citeseer.ist.psu.edu/shaikh02case.html
2. D. Watson, F. Jahanian, and C. Labovitz, "Experiences with monitoring ospf on a regional service provider network," in *Proceedings of the 23rd International Conference on Distributed Computing Systems*, 2003, pp. 204–213.
3. Z. Shu, "Case studies in intra-domain routing instability," presented at the 31th NANOG meeting, May 2004. [Online]. Available: http://www.nanog.org/mtg-0405/shu.html
4. Z. Shu and Y. Kadobayashi, "Troubleshooting on intra-domain routing instability," in *Proceedings ACM SIGCOMM'04 Workshops (NetTs)*, 2004, pp. 289–294.
5. E. Baccelli and R. Rajan, "Monitoring ospf routing," in *Proceedings of IFIP/IEEE International Symposium on Integrated Network Management*, May 2001, pp. 825–838.
6. J. Moy, "OSPF version 2," *RFC 2328*, April 1998.
7. A. Shaikh and A. Greenberg, "OSPF monitoring: Architecture, design, and deployment experience," in *Proceedings of Symposium on Networked Systems Design and Implementation*, March 2004.
8. R. Coltun, D. Ferguson, and J. Moy, "OSPF for IPv6," *RFC2740*, December 1999.
9. D. Oran, "OSI IS-IS intra-domain routing protocol," *RFC1142*, February 1990.
10. RRDtool. [Online]. Available: http://people.ee.ethz.ch/~oetiker/webtools/rrdtool/
11. Intra-domain routing stability measurement project. [Online]. Available: http://rtanaly.koganei.wide.ad.jp/
12. WIDE Project. [Online]. Available: http://www.wide.ad.jp
13. Internet2 abilene backbone network. [Online]. Available: http://abilene.internet2.edu/

Automatic Configuration for VPN Using Active XML

Laurent Ciarletta and Mi-Jung Choi

Madynes, LORIA-INRIA Lorraine, Campus Scientifique - BP 239,
54506 Vandoeuvre-les-Nancy Cedex, France
{laurent.ciarletta,mi-jung.choi}@loria.fr
http://madyne.loria.fr/

Abstract. This paper presents a network management framework for an auto-configuration of dynamic networks using Web Services. The example of VPN configuration is used as a proof of concept, and the Active XML technology provides the underlying Web Services technology with a Peer2Peer infrastructure. We present those technologies applied to network management as well as our prototype implementation, and we discuss the pros and cons of our solution and our future work.

1 Introduction

As Internet and networks grow fast, various network devices are emerging. The configuration of these devices is an essential task specifically regarding changes of network environments. As a part of the SWAN (Self aWAre maNagement) collaborative research project, we are exploring the automatic configuration of dynamic networks. The goal of this project is to study innovative methods for autonomic and automatic management of networks. One of its work subjects is the dynamic configuration and reconfiguration of network elements such as gateways, firewall, probes, etc.

Today's network management architectures are not adapted to the dynamicity of the network elements, their heterogeneity and the scales of future networks. Items such as PDAs or laptops come and go, and the overall dynamicity and needs of data flows inside the network are consistently changing. Users want mobility, ubiquitous connectivity, security, ease of use and quality of service (QoS) while managers want to be able to manage the system at least with a high level view, and to delegate local policies and device specific tasks to subsystems. Even more, P2P architectures and ad-hoc networks are redefining a standard network management model. The standard management information tends to be static (like in MIBs). However, the current network situation is continually changing, so the standard manager/agent model cannot be used 'as is' and the dynamic network management information is required. Therefore, network management needs to be changed to support today's dynamic and diverse networks.

In this paper, we are investigating methods and tools to manage these dynamic networks automatically. We have found Active XML [1], a Web Service framework to fit our needs and it is used as the basis of our work. As the first experiment we are developing a dynamic configuration of a Virtual Private Network (VPN) [2] using Active XML. It will allow the configuration and reconfiguration as well as the monitoring of VPNs using Web Services. While we are primarily focusing on VPNs, i.e. securing the communication, our solution can be applied to QoS as well, where specific bandwidth and critical resources are dynamically associated with security. We

T. Magedanz, E.R.M. Madeira, and P. Dini (Eds.): IPOM 2005, LNCS 3751, pp. 173–180, 2005.

configure VPN endpoints and road-warriors by deploying security policies or device specific configuration files using Active XML. Some parameters can be negotiated between Active XML peers.

The organization of this paper is as follows. Section 2 describes Active XML technology, its usage to network management and its appropriateness as a management solution to dynamic networks. Section 3 explains the design of management framework for dynamic networks and Section 4 illustrates a prototype implementation of our management system using P2P infrastructure. Finally, we conclude our work and discuss directions for future work in Section 5.

2 Active XML

In this section, we briefly explain the Active XML technology that is used in our design and implementations of the auto-configuration system. Also, we show why Active XML is an appropriate solution for management of dynamic networks.

Active XML [1] (or AXML in short) is a declarative framework for distributed data management based on XML and Web services. An Active XML document is an XML document that may contain calls to Web services. That is, AXML is build around AXML documents which are basically XML documents embedding calls (<SC> tags) to other Web Services (that can be AXML or stock Web Services). With this framework, service calls activation and period of validity can be controlled. In some sense, an AXML document can be seen as a (partially) materialized view integrating plain XML data and dynamic data obtained from service calls. It is important to stress that AXML documents are syntactically valid XML documents. As such, they can be stored, processed and displayed using existing tools for XML.

Also, AXML is a peer-to-peer framework for dynamic XML documents in the context of Web Services. An AXML peer is mainly a repository of Active XML documents. By using the P2P paradigm, AXML gives more flexibility and scalability than standard hierarchical architectures: each AXML element or peer can act as a client and a server. The peers can exchange intentional data defined as calls to Web Services. Thus, each Active XML peer acquires and provides dynamic information, in a decentralized peer-to-peer architecture.

We are exploring the use of Active XML as a way to solve some of the management issues related to the high volatility and dynamicity of the network elements and the lack of permanent and generalized infrastructure. We are applying the P2P architecture to network management and proposing a novel management schema.

An AXML peer in the management plane can be a manager and an agent. If the AXML peer is an agent, the peer extracts the management information from the device in the action plane. That is, the management plane provides both management functionalities between managers or between a manager and an agent. The action plane serves a role of acquiring dynamic and actual information from devices. This allows for an increased ease of use and an intrinsic dynamicity. Since AXML peers communicate through Web Services, it can insure interoperability and integration. Since managed data are abstracted, AXML documents can be used on any devices or at a higher level on policy enforcers. Underlying configuration technologies (XML based or not) can be linked with our framework to realize the configuration of any entities.

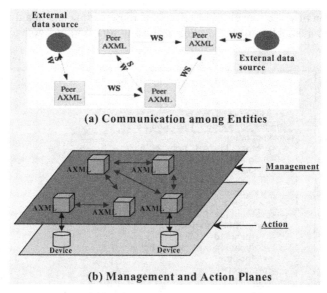

(a) Communication among Entities

(b) Management and Action Planes

Fig. 1. In our management model, we have a P2P architecture consisting of two planes: one is the action plane; the other is the management plane. As shown in Fig. 1 (a), a device in the action plane in Fig. 1 (b) is an external data source. Also, the 'AXML' entity in the management plane is an AXML peer. All these entities communicate with each other through Web Services.

3   Design

In this section, we describe our prototype design details using Web Services. First, we explain the management information for IPsec VPN policies. We also present a sample scenario for VPN configuration.

3.1   Management Information

We build IPsec VPN between our endpoints and dynamically configure the IPsec security [3] related databases (SAD and SPD) using XML documents and Web Services. This means that hosts should be able to auto-configure all their security parameters with taking security policies into account, negotiate and compute those parameters by exchanging information in an XML format and using AXML and Web Services.

Since the standard management information for XML-based IPsec policy does not exist, we have developed our own IPsec-XML document based on CISCO documentation and IETF's documentation [3]. Our document is divided into two sections corresponding to security associations (SAs) and security policies (SPs) parameters.

3.2   VPN Configuration Scenario

We have developed a simple scenario of VPN configuration as a proof of concept for our prototype architecture. In this scenario, mobile users and elements of a corporate network will create specific security parameters at the connection time, allowing the configuration of IPsec tunnels. They negotiate those parameters in a dynamic fashion according to available resource at their activation time and renegotiate them when it is necessary.

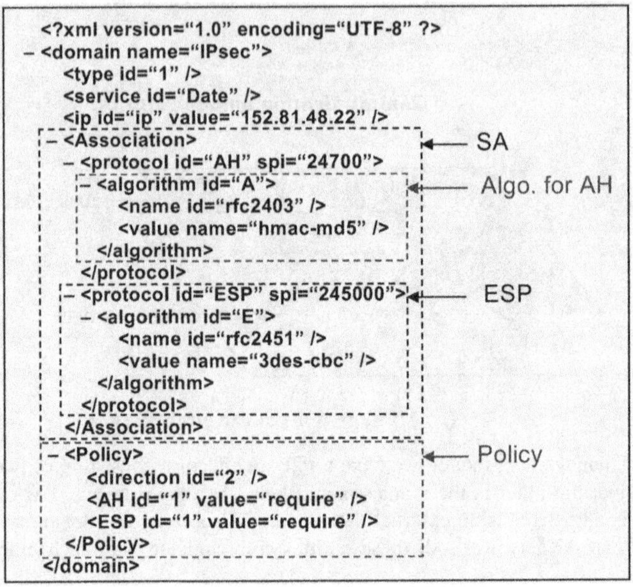

```
<?xml version="1.0" encoding="UTF-8" ?>
- <domain name="IPsec">
    <type id="1" />
    <service id="Date" />
    <ip id="ip" value="152.81.48.22" />
  - <Association>                                    ◄─── SA
    - <protocol id="AH" spi="24700">
      - <algorithm id="A">                           ◄─── Algo. for AH
          <name id="rfc2403" />
          <value name="hmac-md5" />
        </algorithm>
      </protocol>
    - <protocol id="ESP" spi="245000">◄─── ESP
      - <algorithm id="E">
          <name id="rfc2451" />
          <value name="3des-cbc" />
        </algorithm>
      </protocol>
    </Association>
  - <Policy>                                         ◄─── Policy
      <direction id="2" />
      <AH id="1" value="require" />
      <ESP id="1" value="require" />
    </Policy>
  </domain>
```

Fig. 2. This shows an example of an XML document for security policy. This document consists of two parts: SA and SP. This describes simple policies such as 'every connection with a node on a given network needs to use AH and ESP with medium size keys'.

We are building IPsec VPN between our endpoints. IPsec is the IP security protocol and it was initially developed by the IETF to provide network level security for IPv6. It was quickly ported back to IPv4. IPsec systems and routers store low level security parameters and higher-level security policies in dedicated databases. These security policies are high relative to interfaces, not real high level ones. The basic element is a security association (SA) between two hosts storing data regarding security method (AH or ESP) encryption keys and algorithms. IPsec provides 2 main security mechanisms through its two headers ESP (Encapsulation Security Payload, mainly providing confidentiality) and AH (Authentication Header, mainly providing authentication services). The Internet Key Exchange (IKE) [4] allows the automatic exchange/negotiation of IPsec keys.

4 Implementation

The implementation uses the Tomcat servlet engine plus AXIS and Active XML to provide Web Services. The AXML extension provides us with P2P-like behaviors.

Each node also has a security policy document describing simple policies such as 'every connection with a node on a given network needs to use AH and ESP with medium size keys'. Every module is an independent Web Service that could be used by any other Web Service tools. Sub-modules are called in the following sequence:

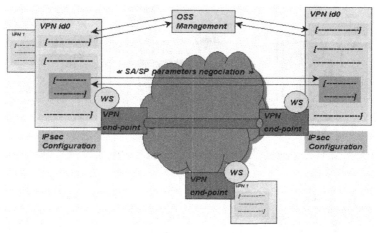

Fig. 3. This illustrates a sample scenario for VPN configuration. First, nodes find the central repository (OSS) where initial configuration and authentication data are stored in the form of generic documents (describing high level requirements for all entities) or device specific documents (describing configuration details for a given node). In the case of a failure to communicate with the repository, the network entities already communicating with other entities can continue to work properly or reorganize. If they are not yet communicating at the time of the failure, but are already linked to other intermediate nodes, they should have a possibility to negotiate through one (or several) trusted entities.

- Step 1: The security module is the global service that calls the security policy modules that can negotiate SA parameters. The role of the security module is to provide authentication of the communicating nodes and secure the following communications by using keys negotiated with a public key algorithm (Diffie-Hellman) [5]. These parameters are thereafter used for the configuration of the secured tunnel for the following AXML data exchange (following steps).
- Step 2: The security policy module negotiates the SAs parameters and defines the security policy for IPsec VPN configuration. This module is the one making and analyzing offers on security policies made by the communicating peers. The corresponding agreements translate into IPsec parameters.
- Step 3: Via the configuration module, we perform the actual configuration with the IPsec parameters on the local node.
- Step 4: The last module (information module) is a monitoring tool that can be queried at run-time and give up-to-date configuration information.

We have developed an installation package for our prototype that works only on Linux. Since we are using standard IPsec configuration tools (setkey, ...), the porting to other Unix OSs should be straightforward. As far as the configuration protocol is concerned, we have decided to implement pre and post check mechanisms. The pre-check insures that we are not enforcing wrong SA parameters on one or both sides, which would lead to a broken link, since IP packets would be tossed by the IPsec security module. The post-check deals with broken-links. In that case, there is a broken/badly-configured link; this mechanism looks for a third party peer with which both elements still have connectivity. The third party element will negotiate new parameters between two peers willing to communicate.

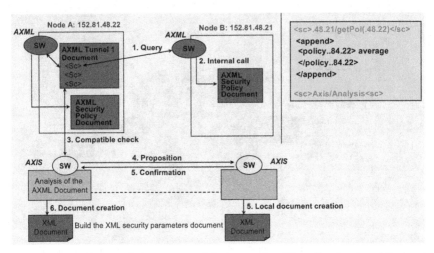

Fig. 4. This shows the negotiation process of 2 given nodes. Given 2 nodes A and B, the security parameters negotiation works as follows:

1. Query: the AXML generic document on node A calls the other AXML engine on node B for querying the security policy.
2. Internal call: node B asks for its own security policies and builds up the AXML document with these pieces of information.
3. Compatibility check: a third call is sent to the AXML document analysis Web Service that checks whether two policies are compatible or not.
4. Proposition: then, it makes an offer to the distant peer.
5. Confirmation & Local documentation creation: if the offer is validated by both entities, local XML configuration documents are created.
6. Document creation: when the XML documents have been created they are used to populate both SAD and SPD on each node.

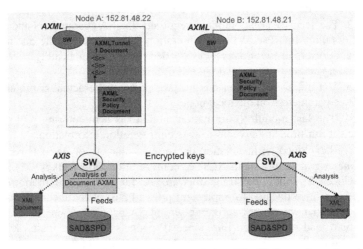

Fig. 5. This shows the results after finishing negotiation. The XML configuration documents that have been produced are processed to check their validity for IPsec configuration and then used to configure the IPsec parameters of the node (SAD and SPD).

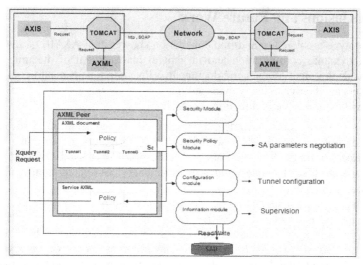

Fig. 6. This shows the implementation architecture for our prototype. Each AXML peer has a repository of XML documents for VPN configuration. These documents are at first retrieved from a central repository (OSS). Each document corresponds to a tunnel (a bi-directional tunnel between a peer and another one). Each document embeds 4 sub-services (or modules): security module, security policy module, configuration module, and information module.

We have also developed a complete configuration interface. Each node can be manually controlled with a distant Web Service, and performances can also be measured through a packet-sniffing module integrated into the supervision console. This last tool was developed to compare performances between our full Web Service-based configuration and a mixed-mode IKE/AXML configuration schema where we used 'racoon' [6], an open-source implementation of IKE [4] to do the key exchange mechanism. We studied the activation date for the security parameters (encryption keys). Our active XML solution is roughly two times slower than the IKE mechanism in the exchange of security parameters between 2 nodes.

These experiments done between two machines with several network interfaces showed that the use of Web Service could not beat the native IKE mechanism as far as speed is concerned. AXML cannot work in scenarios where keys should be created or renewed too often (under several seconds), but our system has the advantage of separating parameters negotiation from their instantiation. This means that these negotiations can be planned, and in the event of concurrent renegotiation, they could be delayed and be reordered to avoid network flooding of IKE packets for example. They could also be combined with other configuration data.

Even though we are proposing a key exchange mechanism, our solution is not a replacement for IKE but a proof of concept. Since it is based on XML technology, and on Web Services, its performances are not supposed to be better than a dedicated technology such as IKE. However, Web Services technologies such as AXML are generally not affected by firewall issues because it uses HTTP ports. When configuration is considered as a whole (not only for IPsec parameters), then our generalized framework allows multiple configurations at the same time with one generic and consistent tool.

5 Conclusions and Future Work

Our prototype architecture and implementation showed that AXML is a good candidate for dynamic configuration and in general management of dynamic networks. After the deployment of a generic document, VPN endpoints can auto-configure their parameters and be dynamically managed.

We will explore several directions in our future work:

Firstly, we will look into integrating standards in our solution at several levels, from the configuration module with Network Configuration (NetConf) [7] to the policy aspects with the Security Policy Specification Language (SPSL) [8] improving our proprietary IPsec language, and adding Web Services related security protocols to secure our framework.

Secondly, the implementation has not been thoroughly tested as far as its behavior and dynamicity is concerned. AXML creates a chain of Web Service calls, therefore consistency and synchronization issues need to be investigated. These are closely related to Web Services orchestration or choreography. Also, we need to examine performance and scalability issues through the implementation and the testing.

Finally, we need to provide a complete auto-configuration and reconfiguration solution. We are looking into strengthening our framework when dealing with failure and inconsistency, and removing the need for a permanent central repository. We are also searching ways to automatically discover the management entities, capabilities of network elements.

References

1. S. Abiteboul, O. Benjelloun, I. Manolescu, T. Milo and R. Weber: Active XML: a data-centric perspective on Web Services. Evry, France, October 2002
2. Virtual Private Network Consortium (VPNC). http://www.vpnc.org
3. S. Kent, R. Atkinson: Security Architecture for the Internet Protocol. IETF, RFC 2401, November 1998
4. D. Harkins, D. Carrel: The Internet Key Exchange (IKE). IETF, RFC 2409, November 1998.
5. W. Diffie and M.E. Hellman: RSA Data Security, Inc. Public-Key Cryptography Standards (PKCS). IEEE transactions on Information Theory, 1976, pp. 644-654
6. IPsec-Tools. http://ipsec-tools.sourceforge.net
7. IETF: Network Configuration (Netconf).
   http://www.ietf.org/html.charters/netconf-charter.html
8. M. Condell, C. Lynn, J. Zao: Security Policy Specification Language (SPSL). IETF, Internet Draft, http://www.ietf.org/proceedings/00jul/I-D/ipsp-spsl-00.txt

Evaluation of the Fast Handover Implementation for Mobile IPv6 in a Real Testbed

Albert Cabellos-Aparicio*, Jose Núñez-Martínez, Hector Julian-Bertomeu,
Loránd Jakab, René Serral-Gracià, and Jordi Domingo-Pascual

Universitat Politècnica de Catalunya,
Departament d'Arquitectura de Computadors, Spain
{acabello,jnunyez,bertomeu,ljakab,rserral,jordid}@ac.upc.edu

Abstract. Fast Handovers is an enhancement to the Mobile IPv6 protocol, currently specified in an IETF draft, which reduces the handover latency. This can be beneficial to real-time applications. This paper presents a novel implementation of Fast Handovers and an analysis of the handover. Using a real testbed we study the handover latency and the provided QoS: analyzing the OWD, IPDV and Packet Loss before and after the handover. Finally we present a comparison between the Mobile IPv6 and the Fast Handovers handover.

1 Introduction

In the past years Wireless LAN (IEEE 802.11) [1] has evolved and become cheaper considerably. A great interest exists among users in being on-line without wires. In current Internet status, a user can be connected through a wireless link, but he cannot move. That's why IETF designed Mobile IP. This protocol, jointly with WLAN is able to provide mobility to the Internet. In other words, a wireless user with Mobile IP can move from one point of attachment to another without losing the network connections. That's because it will have a fixed IP address that will not change regardless of the location. The most critical part of this technology (WLAN + Mobile IP) is the handover. During this phase, the mobile node (MN) is not able to send or receive data, and some packets may be lost or delayed due to intermediate buffers. This is often unacceptable for real-time or streaming applications (i.e. VoIP).

According to the measurements performed in [2], the WLAN/IPv6/Mobile IPv6 handover takes about 2 seconds. This time is unacceptable for VoIP traffic. The IETF "MIPv6 Signaling and Handoff Optimization" working group has designed Fast Handovers for Mobile IPv6 (FMIPv6) [3] in order to speed it up. Fast Handovers' main goal is to reduce both the handover latency (the duration of the handover) and the packet losses to zero.

This paper presents a novel and unique implementation of Fast Handovers. Our implementation runs on Linux and, as far as we know, is the first public

---

* This work was partially funded by the MCyT (Spanish Ministry of Science and Technology) under contract FEDER-TIC2002-04531-C04-02 and the CIRIT (Catalan Research Council) under contract 2001-SGR00226

T. Magedanz, E.R.M. Madeira, and P. Dini (Eds.): IPOM 2005, LNCS 3751, pp. 181–190, 2005.

implementation of the FMIPv6 protocol [17]. Our goal is to study the FMIPv6 handover in a real testbed using passive and active measurements. We aim to study the handover latency and the provided QoS level analyzing important parameters such as One-Way-Delay, Inter Packet Delay Variation and Packet Loss. We apply the methodology explained in [2] to evaluate the performance of the protocol. We also compare the QoS parameters after and before the handover, we study the differences between the Mobile IPv6 and the FMIPv6 handover and finally, we evaluate if, as stated in [3] FMIPv6 is suitable for VoIP.

Several papers focus on the handover measurement, [4] studies the handovers of different mobility protocols using a simulator, [5] studies the handover latency without taking into account the wireless handover through a mathematical model, [6] studies the FMIPv6 performance through a simulator, [7] proposes a new algorithm to improve the handover latency of the WLAN/Mobile IPv6 handover and finally, [8] makes an empirical analysis in the 802.11 handover. Our paper goes further, analyzing a real implementation of the protocol in a testbed, studying the overall performance of the protocol, especially during the handover and comparing it with Mobile IPv6.

2   Wireless and Mobility Protocols Overview

2.1   IEEE 802.11

The Wireless LAN protocol [1] is based on a cellular architecture, where each cell is managed by a Base Station (BS, commonly known as Access Point or AP). Such a cell with the BS and the stations (STA) is called a Basic Service Set (BSS) and can be connected via a backbone (called Distribution System or DS) to other cells, forming an Extended Service Set (ESS). All these elements together are one single layer 2 entity from the upper OSI layers' point of view. APs announce their presence using periodic "Beacon Frames" containing synchronization information. If a STA desires to join a cell, it can use passive scanning, where it waits to receive a "Beacon Frame" or active scanning, when it sends "Probe Request" frames and receives a "Probe Response" frame from all available APs. Scanning is followed by the Authentication Process and if that is successful, the Association Process. Only after this phase is complete the STA capable of sending and receiving data frames. STAs are capable of roaming, i.e. moving from one cell to another without loosing connectivity but the standard does not define how it should be performed, it only provides the basic tools for that: active/passive scanning, re-authentication and re-association.

2.2   Mobile IPv6

Mobile IP was designed in two versions, Mobile IPv4 [9] and Mobile IPv6 (MIPv6) [10]. The protocol's main goal is to allow MNs to change its point of attachment to the Internet while maintaining its network connections. In other words, the mobile node has a special IP address (Home Address or HAd) that will remain unchanged regardless of the MN's location, moreover, the MN will use temporary IP Addresses (Care-of-Address or CoA) when connected to foreign networks (not its home network), however, it is still reachable through its HAd (using tunnels or with special options in the IPv6 header). A special entity (Home Agent or HA) manages MN's localization by binding the MN's CoA to MN's HAd.

MIPv6 has three functional entities: the Mobile Node (MN) which is any mobile device with a wireless card and the MIPv6 protocol, the Home Agent (HA) which manages MN's localization and finally the Correspondent Node (CN), a fixed or mobile node that exchanges data packets with the MN.

The protocol has four phases. Initially in the Agent Discovery phase the MN has to discover if it is connected to its home network or to a foreign one. IPv6 routers send periodically "Router Advertisements" including network prefix information. The MN will listen to those messages discovering at which network it is attached and will obtain a CoA if it is not in the home network. Next, in the Registration phase, the MN must register its CoA (where it is located) to the HA and CNs in order that they can bind it with the HAd. After this phase, Registration and Tunneling comes, the MN establishes tunnels (if necessary) with the HA and CNs in order to send or receive data packets. Notice that the CNs will still send packets to the same destination IP address (the HAd). The last phase is the Handover, the MN changes its point of attachment and it must discover in which network it is connected once again (Agent Discovery) and register its new CoA (Registration). During this phase some data packets can be lost or delayed due to incorrect MN location.

2.3 Fast Handovers

FMIPv6 is a MIPv6 handover enhancement that reduces the handover latency and stores packets delaying them instead of losing them. This is accomplished by allowing the MN to send packets as soon as it detects a new subnet link (IEEE 802.11 in our case) and delivering packets to the MN as soon as its attachment is detected by the new access router.

FMIPv6 has different operational procedures, for instance, in the "Predictive Handover" the MN discovers nearby APs using the IEEE 802.11 "scan" and then requesting all the important information related to the corresponding new access router. When attachment to an AP takes place, the MN knows the corresponding new router's coordinates including its prefix, IP address and MAC address. Through special "Fast Binding Update" and "Fast Binding Acknowledgment" messages the MN is able to formulate a prospective new CoA (without changing its point of attachment), this CoA must be accepted by the new access router prior to the MN movement. Once the MN has changed its point of attachment and it is connected to the new access router link, it can use its new CoA without having to discover the subnet prefix, it also knows the new access router MAC and IPv6 address, and hence this latency is eliminated. As soon as it is attached the MN sends a "Fast Neighbor Advertisement" announcing its presence. Moreover, the previous access router will tunnel and forward packets to the new care of address until the MN sends a "Binding Update" registering its new CoA to HA and CNs, hence, any packet is lost. The other FMIPv6 operational procedure is the "Reactive Handover" which is very similar to the previous one, however this is not supported by our implementation.

3   Fast Handovers Implementation

3.1   Overview

Our FMIPv6 implementation is written in C and runs on Linux Kernel 2.4.26, it enhances the Mobile IPv6 MIPL 1.1 [11] implementation and complies with the draft–ietf–mipshop–fast–mipv6-03.txt. The basics parts of the draft are implemented, some optional and error recovery parts are under development as future work. However the non-implemented parts do not affect the performance of the protocol, which is the paper's main goal. Our implementation also supports any wireless card (with Linux support) through the "Wireless Tools for Linux" [12].

3.2   Implementation Structure

This section describes the FMIPv6 implementation structure which is mainly divided into two modules:

- *fh–base:* This is a "dumb" module that runs into the kernel and interacts with the IPv6 module, the MIPL module and Netfilter. It receives commands from the user space.
- *fh–daemon:* This is a user-space daemon, interacts with the user, the wireless interface (through netlink) and actually implements the FMIPv6 protocol. It communicates with the "dumb" fh–base kernel module to perform the protocol operational procedures.

We have splitted the implementation into two parts, user–space and kernel–space. The FMIPv6 protocol is an ongoing work and, we can easily adapt the fh–daemon (running on user-space) without having to change the kernel part (the most difficult one).

3.3   Development Environment

Developing support for a new protocol for the Linux Kernel is not an easy task, especially if it has to interact with other modules. In order to have a productive development environment we used User-Mode-Linux (UML) [13]. UML provides a virtual machine that emulates a Linux Box. We recreated our real testbed using UML on a single physical machine, all the virtual machines had the same configuration than the real ones, we used the same network topology, the same kernel and software versions. IEEE 802.11 is not supported by UML, however we emulated the handover using IEEE 802.1 and we simulated movement between two switches. The IEEE 802.11 part of the implementation was only tested in the real testbed. With this development environment we were able to intensively test our implementation in an easy and affordable way. Only after the implementation was mature enough, we moved it to the actual testbed to test it and to measure the FMIPv6 handover.

4   Measurement Scenario

The testbed's main goal is to test the FMIPv6 implementation in a real scenario, evaluate its handover latency using passive measurements and measure the important QoS parameters using active measurements. The testbed is shown in Figure 1, all the machines are synchronized using NTP (Network Time Protocol) obtaining 1ms accuracy. See [2] for further details.

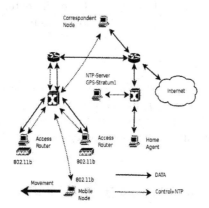

Fig. 1. Simplified measurement scenario

All the machines belonging to the testbed are using the GNU/Linux Debian Sid distribution, however the hardware depends on the role of each computer. Both access routers have two wireless cards, one for communicating with the MN and the other one to capture frames (passive measurements). Those cards have the Atheros Chipset (802.11b). The running kernel is 2.4.26 patched with the FMIPv6 implementation. The MN uses a wireless Cisco Aironet 350 card, the running kernel is also 2.4.26 patched with MIPL 1.1 and with the FMIPv6 implementation. Finally the HA and the CN use the 2.4.26 kernel patched with the MIPv6 software (there is no need of FMIPv6).

5   Methodology

We will apply the methodology depicted in [2] to evaluate the FMIPv6 handover.

5.1   Passive Measurements

The handover latency is the time spent during the handover. To compute it we developed a special tool "PHM" that monitors the signaling messages in both APs of our testbed. We capture all the packets sent or received by the wireless interface using Ethereal. The handovers are forced using special user-space wireless utilities for Linux [12]. When the MN has received the "Fast Binding

Acknowledgment" message it is ready to move to the new access router. At that point we force the wireless card to change from the old AP to the new one. Our FMIPv6 behaves as stated in [14]. As soon as our implementation detects the new link (using [12] once again) we send the "Fast Neighbor Advertisment" to announce the MN presence. Once the handover is finished and having the frames captured by Ethereal, PHM processes the signaling messages off-line providing the computation of the handover latency. Moreover PHM is able to differentiate between the different parts of the handover latency (Scanning, Authentication and Association for 802.11b). In fact, PHM is easily extensible to other mobility protocols and is able to compute the handover latency also for MIPv6.

5.2   Active Measurements

Using active measurements we intend to analyze the provided QoS at IP level. The basis of such tests is to generate a synthetic flow traveling through the network under test. The developed application to make such measurements is NetMeter [15] and we apply the methodology presented in [2] to perform the active measurements.

5.3   Evaluation of the FMIPv6 Implementation

For a good analysis of the handover, it is necessary to build up a good set of tests. In this paper we ran a set of 10 tests each 5 minutes long, from where we extracted a set of 40 valid handovers.

In order to evaluate the protocol and our implementation we used two different packet rates and sizes. Half of the tests had 64kbps traffic. This flow simulates with UDP the properties of VoIP traffic under IPv6. It sends 34 packets per second with 252 bytes of payload as stated in [16]. Due to the low rate needed for VoIP the other tests are done on a higher packet rate, so the impact of a different bandwidth can be studied. This flow (Data) sends 84 packets per second with a payload size of 762 bytes per packet. The paper's main goal is to analyze our FMIPv6 implementation, check if it works as expected and provide performance results, especially regarding its handover latency and the QoS parameters. To test our implementation under stress conditions requires having multiple MNs and APs which is very difficult to deploy in a real testbed. These kind of tests are left as future work and will be done using the UML infrastructure.

All the tests are from the CN to the MN. With FMIPv6, when the packets flow in this direction, the access routers must tunnel and buffer packets showing an interesting behavior. However, when the traffic source is the MN, there is no need to tunnel packets, just to buffer them on the MN (the FMIPv6 handover latency remains constant for both directions), that's why we focus on the CN–>MN direction.

6   Results

This section describes the results obtained from the tests discussed in the previous section.

6.1   Handover Latency

Figure 2 shows an instantaneous One-Way-Delay (obtained with NetMeter) where it is easy to see the handover. We can see that no packet is lost; regarding the delay we see a spike. This behavior is due to FMIPv6, while the MN is changing its point of attachment (from one AP to the other) the old access router is tunneling and forwarding packets to the new access router and the new access router, at the same time, it is buffering packets until the MN regains connectivity. So, FMIPv6 delays (buffers) packets instead of loosing them. The packets will be stored in a buffer while the MN's WLAN layer is disconnected; hence, this delay is equal to the 802.11 handover latency (see the numerical results).

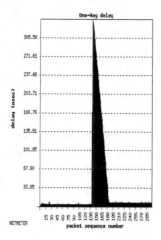

Fig. 2. FMIPv6 handover, instantaneous OWD (VoIP Traffic)

Once the handover is finished we can see that the delay is slightly higher than before, that's because the packets are being routed to the old access router and tunneled to the new access router, introducing an extra hop. When the MN sends a "Binding Update" to its HA and CNs the traffic will be routed directly to the MN. However the MIPL Mobile IPv6 implementation does not support this enhancement and it is not implemented. Table 1 shows the results obtained with our PHM application and are the numerical results of the FMIPv6 handover latency (results are in milliseconds).

Table 1. FMIPv6 handover latency (ms)

	Mean	Std.Dev.
VoIP Traffic	319.05	25.67
Data Traffic	330.34	29.22

The PHM tool shows that the FMIPv6 handover latency is equal to the IEEE 802.11 handover latency (as expected) computed in [2] and [5]. The rate and the packet size do not affect the handover latency.

6.2 QoS Parameters Analysis

Table 2 summarizes all the results obtained with NetMeter regarding the provided QoS level of the FMIPv6 handover. This is accomplished by taking 100 packets before the handover and calculating the OWD, the IPDV and the same after it.

Table 2. FMIPv6 One-Way-Delay and Inter Packet Delay Variation (ms)

		OWD (ms)		**IPDV** (ms)	
		Before	After	Before	After
VoIP Traffic	Mean	2.77	6.17	16.68	31.4
	Std.Dev.	1.40	4.52	16.4	37.0
Data Traffic	Mean	7.27	17.3	16.54	63.9
	Std.Dev.	2.73	17.1	22.41	65.0

These numerical results confirm that the delay is slightly higher after the handover due to the extra hop. They also show that the OWD remains constant before and after the handover for VoIP traffic. For longer packets (762 bytes) the OWD has variance after the handover (17ms of IPDV after the handover). [2] shows important QoS fluctuations in the Mobile IPv6 handover due to the wireless card. In a MIPv6 handover the wireless card decides to switch to a new access point regardless of the above layers, it changes its point of attachment when it detects a signal degradation [1], hence, the provided QoS is severely affected, especially for longer packets. In FMIPv6 the wireless card is forced (by the above layers) to switch from one AP to another one without having to wait until the signal degrades. The FMIPv6 OWD variance for long packets after the handover may be due to implementations issues, packets must be tunneled and forwarded, not just forwarded. However, the results provided in [2] shows that MIPv6 suffers from a higher variance than FMIPv6.

6.3 Fast Handovers vs. Mobile IPv6

Figure 3 shows a Mobile IPv6 handover [2] where we can clearly see the gap produced by the handover.

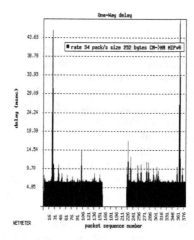

Fig. 3. MIPv6 handover, instantaneous OWD

During this period of time, the MN is not able to send or receive data, thus the packets are lost, while the FMIPv6 implementation does not lose any packet. Regarding the OWD/IPDV before and after the handover, as explained above, Mobile IPv6 suffers from a higher IPDV before the handover due to the decrease of the signal strength (especially for long packets). Finally, regarding the handover latency (the time for interruption), Mobile IPv6 has approximately 2 seconds [2] while in FMIPv6 is about 325ms. This handover latency produces packet losses in Mobile IPv6 that may be computed as the rate multiplied by the handover latency.

7   Conclusions

This paper presents a novel Fast Handovers implementation and analyzes through active and passive measurements the protocol handover in a real testbed. The analysis focuses on the handover latency and the level of provided QoS (OWD, IPDV and PL). Finally it compares the performance obtained between Mobile IPv6 and Fast Handovers in a real testbed.

The results obtained through passive measurements show that the FMIPv6 handover latency is equal to the WLAN handover latency, therefore, FMIPv6 reduces the IPv6 and MIPv6 handover latency to zero and is as fast as the WLAN handover. Active measurements show that, in FMIPv6 there is a light QoS degradation after the handover for long packets, whereas in MIPv6 the WLAN signal strength degrades and there is a severe OWD variance. Moreover, while MIPv6 loses packets, FMIPv6 delays them. In the worst case a packet is delayed a 'WLAN handover latency' (about 325ms) which is often acceptable for VoIP traffic.

The FMIPv6 protocol and our implementation achieve the expected goals. In [17] are the FMIPv6 implementation NetMeter and PHM Tool available under the GPL license. Also all the detailed results and several figures are available.

References

1. IEEE 802.11: Wireless LAN Medium Access Control and Physical Layer. (1997)
2. A. Cabellos-Aparicio, R. Serral-Gracià, L. Jakab, J. Domingo-Pascual: Measurement based analysis of the handover in a WLAN MIPv6 scenario Passive and Active Measurements 2005, Springer, LNCS 3431, pp 203-214
3. R. Kookl: Fast Handovers for Mobile IPv6 draft-ietf-mipshop-fast-mipv6-03.txt
4. N. Montavont and T. Noel: Handover Management for Mobile Nodes in IPv6 IEEE Communications Magazine (2002)
5. A. Mishra, M. Shin and W. Arbaugh: An Empirical Analysis of the IEEE 802.11 MAC Layer Handoff Process Vol.33 ACM SIGCOMM Comp. Comm. Review
6. Marc Torrent-Moreno, Xavier Pérez-Costa, Sebastià Sallent-Ribes: A Performance Study of Fast Handovers for Mobile IPv6 28th IEEE Inter. Conf. on Comp. Networks
7. Marco Liebesch, Xavier Pérez Costa and Ralph Scmitz: A MIPv6, FMIPv6 and HMIPv6 Handover Latency study: Analytical Approach IST Mobile and Wireless Telecommunications Summit (2002)
8. Xavier Pérez Costa and Hannes Hartenstein: A simulation study on the performance of Mobile IPv6 in a WLAN-based cellular network Volume 40 issue I Computer Networks: International Journal of Telecommunications Networking (2002)
9. C. Perkins: IP Mobility Support for IPv4 RFC 3344 (2002)
10. D. Johnson, C. Perkins and J. Arkko: IP Mobility Support for IPv6 RFC 3775
11. HUT: MIPL Mobile IPv6 for Linux (online) *http://www.mobile-ipv6.org/* (2004)
12. Wireless Tools for Linux (online)
    http://www.hpl.hp.com/Jean_Tourrilhes/Linux/Tools.html
13. Jeff Dike: UML – User-Mode-Linux (online) *http://user-mode-linux.sourceforge.net*
14. P. McCann: Mobile IPv6 Fast Handovers for 802.11 Networks draft-ietf-mipshop-80211fh-04.txt
15. R. Serral-Gracià, A. Cabellos-Aparicio, H. Julian-Bertomeu, J. Domingo-Pascual: Active measurement tool for the EuQoS Project. IPS-MOME 2005 Workshop.
16. John Q. Walker, NetIQ Corp: A Handbook for Successful VoIP Deployment Network Testing, QoS and More (2002)
17. SAM - Advanced Mobile Services - Research Project MCyT (online) *http://sam.ccaba.upc.edu*

A Trial Experience on Management of MPLS-Based Multiservice Networks

Eduardo Grampín[1], Javier Baliosian[2], Joan Serrat[3],
Gonzalo Tejera[1], Federico Rodríguez[1], and Carlos Martínez[1]

[1] Universidad de la República, Montevideo, Uruguay
{grampin,gtejera,rodrigue,carlosm}@fing.edu.uy
[2] Network Management Research Centre, Ericsson Ireland
javier.baliosian@ericsson.com
[3] Universitat Politècnica de Catalunya, Barcelona, Spain
serrat@tsc.upc.edu

Abstract. This article presents a component-based, distributed management system for Multiprotocol Label Switched (MPLS) multiservice networks[1]. Delivery of "triple play" multimedia services to the broadband residential user is a demanding challenge. The complexity is increased by the requirement of preserving Quality of Service (QoS) assurance for legacy connectivity services to the enterprise segment over the same infrastructure. New technologies are being introduced in the access, aggregation and core networks. Management applications must be aware of these advances and shall evolve accordingly. The proposed management architecture benefits from the capabilities of the MPLS Control Plane, in conjunction with a traditional management approach to provision QoS-aware services. This hybrid solution pursues short connectivity setup times by means of Control Plane signalling, with Traffic Engineering capabilities provided by the management framework. The system is being prototyped on a trial metropolitan testbed. Simulation results show that an advantageous trade-off between speed and resource optimisation is feasible.

1 Introduction

The convergence of video, voice and data services over the Internet occurs in parallel with the convergence of the historical Internet and Telecommunications management frameworks into object-oriented, distributed management paradigms. Convergent multiservice networks are build over several technology layers that must be tied up together in order to provide "triple-play" services to the broadband end user. ADSL "first mile" user access, ATM, Gigabit Ethernet and/or SDH aggregation and core networks with proper dimensioning and Traffic Engineering techniques enable the delivery of voice, Interactive TV and High Speed Internet services. Furthermore, these networks must also support, among others, traditional best effort Internet and VPN services for the Corporative segment, augmented with Datacenter (IT) services. The overwhelming operational complexity of such multiservice networks requires the

---

[1] This work is part of a project undertaken by Universidad de la República (UdelaR) and ANTEL, the public Telco of Uruguay, for development of a Metropolitan Multiservice Network with MPLS over optical transport, funded by Programa de Desarrollo Tecnológico (PDT)

T. Magedanz, E.R.M. Madeira, and P. Dini (Eds.): IPOM 2005, LNCS 3751, pp. 191–201, 2005.

appropriate integration at the Data, Control and Management Planes. Traditional layered non-integrated systems are simply unable to cope with the multiservice demand. IP/MPLS can provide some degree of integration at the Data and Control Plane, while modern management paradigms can complement such integration at the Management Plane.

This paper presents the preliminary results obtained by an ongoing project aimed at the study of the technological, operational and management problems arising in the deployment of a multiservice MPLS-based network. A component-based, distributed management system named GERMINA, which stands for "GEstor de Red del grupo MINA" [1] (MINA research group Network Management System), is being developed to face the O&M challenges. GERMINA comprises the Element and Network Management Layers (EML, NML), with a light Service Management interface that enables future development of the top layers of the traditional ITU-T TMN hierarchy.

The paper is structured as follows. Section 2 describes the GERMINA management system, while Section 3 reviews relevant related work. Section 4 presents early evaluation results, and the last section is devoted to make some concluding remarks and to point out future work.

2 GERMINA Management System

The evolution path towards the next generation IP over optical networks is supported by the MPLS Control Plane, which provides functionality to solve path computation and path establishment (i.e., the provisioning process); however, there are alternative proposals to support connectivity provisioning based on the Management Plane. For instance, the IST WINMAN project specified an Integrated Network Management System (INMS) for providing IP over WDM connectivity services, mostly using management functions supported by the Control Plane wherever applicable, as presented in [2].

Other management systems presented in Section 3, confirm that the intelligent Control Plane provides an important part of the operational solution in MPLS networks, which shall be complemented with well-known, trusted management techniques and tools. In this regard, a set of definitions and tools that provide support for Operation and Management (OAM) of MPLS networks are becoming available [3].

GERMINA is founded over JacORB [4], which provides the framework for the integration of components, developed by different groups of people spread in time and space. The Network and Element Management (NML-EML) interface is modelled following the Multi Technology Network Management (MTNM) Information Model [5], augmented with the MPLS extensions developed by the aforementioned IST WINMAN project.

2.1 Scenarios

This section presents some representative scenarios in order to state the general principles of GERMINA system operation to justify the system architecture depicted in Fig. 1. , which will be described later in Section 2.2:

- Service Provisioning
- SLA surveillance

- Offline Reoptimisation
- LSP Provisioning triggered by the Control Plane

Service Provisioning

Service Provisioning is triggered by the Service Manager (SM), as a result of the interaction with customer care applications, which defines service attributes, including specific Service Level Agreement (SLA). This SLA is translated by the SM into a Service Level Specification (SLS) in order to be implemented by the provisioning processes. Once the service specification is complete, a provisioning request is issued to the Configuration and Performance Manager (CPM), which proceeds with the phase of resource reservation and connectivity setup.

In order to fulfil the requested Quality of Service (QoS) it is necessary to solve a Constraint-Based Routing (CBR) problem for each request. This process is performed by the Routing and Management component (RMA) upon request of the CPM. The RMA has detailed knowledge of network topology, which is gathered running the Internal Gateway Protocol (IGP) as a peer on the network. The resources selected to satisfy the ongoing connectivity request are reserved in the internal CPM Inventory Component.

The provisioning process progresses in the Management Plane side towards the Element Manager (EM), which configures the required SubNetwork Connections (SNCs) over the technology neutral managed objects. This configuration is translated into technology/device specific commands to implement the changes in the network by means of a suitable protocol such as SNMP and/or Command Line Interface (CLI).

Finally, once the changes have been implemented, the upper layers of the management framework become aware of the topology changes by CORBA notifications issued by the EM (i.e. the Inventory is updated by these notifications).

SLA Surveillance

The Performance and Fault Management processes of the EM trigger an SLA surveillance process when given thresholds are reached in relevant parameters and/or faults are detected in network elements. Several mechanisms can be used to detect these anomalies, depending on equipment peculiarities. Most vendors implement SNMP Agents embedded in network devices with trap generation capabilities under certain faulty conditions and/or when thresholds are reached in relevant objects (i.e. percentage of link bandwidth usage). Other possibilities include external monitoring and measuring by dedicated appliances (e.g. RMON probes).

Performance degradation is detected at the Network Management Layer by means of notifications issued by the EM. On arrival, the Performance Manager first consolidates the information, and then checks the conformance of configured SLSs. If a service violation is detected the CPM may react in several ways; if (1) a set of policies are defined, the CPM will initiate a reconfiguration process in order to meet the SLS parameters, else (2) the upper management layer (SM) will be notified in order to proceed with a service re-establishment. A complete implementation shall also trigger commercial processes such as discount and report to the affected client(s).

Offline Re-optimisation

The objective of this process, which can be periodically scheduled by network administrators (e.g. once a week), is to re-compute the optimum resource usage for a known traffic demand. Given an optimum distribution of resources at a given time (e.g. after offline re-optimisation), network states evolve out of the optimum due to dynamic resource assignment. The CBR process computes feasible resource distribution for each request, but in general a global optimum is not reachable under this dynamic demand regime.

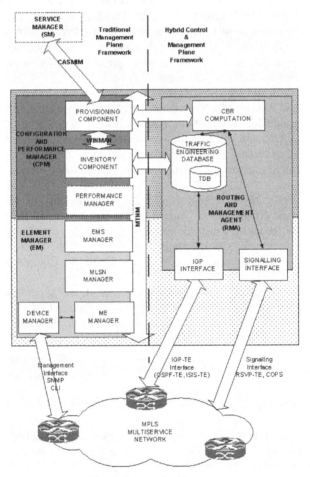

Fig. 1. System Architecture

Offline re-optimisation is performed using a snapshot of the Topology Database and taking into account global optimisation objectives, which are defined by network administrators in the form of objective functions and administrative constraints (i.e. policies), considered by the CBR algorithm. Once the global optimum is computed by the RMA, the system shall reconfigure network connectivity services. Care must be taken to avoid service disruption, using a "make before break" procedure.

LSP Provisioning Triggered by the Control Plane

The *Service Provisioning* scenario described a traditional Management Plane-initiated provisioning process. GERMINA is capable of LSP setup upon request of an ingress

LSR. The request is received by the RMA using its *Signalling Interface*. There are two optional approaches for signalling, namely the request/reply transaction between the RMA and the ingress LSRs, as described in Section 2.2.3. Once a path has been computed for a particular request, the ingress LSR receives a reply from the RMA and LSP setup proceeds using the standard Control Plane signalling.

An interesting feature of this approach is that the RMA can compute both the downstream and the upstream paths. Thus, bidirectional LSP setup is enabled, triggering standard signalling in both the ingress and the egress LSR.

2.2 Description of System Components

The system implements basic Element (EML) and Network Management (NML) layers. The existing functionality comprises Configuration, Performance and limited Fault Management (basically provided by the EML).

Note that the left side of the Fig. 1. presents a "traditional" distributed, component-based management framework founded over the mentioned CORBA bus, while the right side shows the RMA modules, which implement the aforementioned hybrid connectivity setup approach. The RMA management interfaces (through the CBR and Topology DataBase modules) enable the interaction with the rest of the framework for the fulfilment of the functionality needed in the above described scenarios. The hybrid approach can work in an autonomous fashion and will be described in Section 2.2.3. The following sections describe the GERMINA components in detail.

2.2.1 Service Manager

This module has been partially implemented as a proof of concept of the NML and EML functionality, using a minimal version of the CASMIM SML-NML interface [6]. The SM invokes methods provided by the NML related to Subnetwork Connection lifecycle:

- createSNC();
- activate SNC();
- deactivateSNC();
- deleteSNC();

2.2.2 Configuration and Performance Manager

This module is driven by a *Provisioning Component* (PC) which makes use of the *Inventory Component* (IC) and the *Routing and Management Agent* (RMA) components to fulfil the Configuration Management processes. PC and IC have been developed by coordinated BSc Projects [7][8], while the *Performance Manager* (PM) component is partially developed, aiming to handle network-level performance policies using the functionality provided by the EM and under the surveillance of the SM objectives. Note that PC and IC have an additional interface that uses the Information Model developed by the IST WINMAN project for object persistency management.

2.2.3 Routing and Management Agent

The RMA is a logically centralized entity intended to perform computation of network paths under a given topology with QoS and administrative constraints, and without assuming previous knowledge of future demand. In other words, the RMA is a Constraint Based Routing (CBR) capable entity, and behaves like a Path Computation Server (PCS) with the following additional and innovative features:

- The RMA gathers network topological information as an IGP peer in a given domain.
- Path computation requests can be signalled by ingress LSRs using the standard RSVP-TE signalling protocol [9], or using a request/reply protocol, as being defined by the IETF Path Computation Element (PCE) Working Group [10].
- Path signalling is performed by ingress LSRs using the standard RSVP-TE signalling protocol.
- The RMA can be integrated in a management distributed component-based environment using well-known frameworks such as CORBA, J2EE, .NET, etc.
- The RMA hardware and software architecture is a distributed computational system based on High Performance Computing (HPC) principles, with virtually unbounded resources.

Traditional *offline CBR* performs path computation outside network elements, in a PCS. Such process takes as input a known, static traffic matrix (result of existing connections and connectivity requests) and, based on a detailed and accurate topology map (built with information gathered from the network), computes the *optimal* network paths for that given traffic matrix.

On the other hand, *online CBR* is a routing mechanism embedded on the network elements intelligence. Such a routing process receives as input dynamic traffic requests and has no knowledge of future requests. Given this traffic request and based on dynamic (and possibly incomplete) network state it computes a *feasible* path for that request.

CBR was traditionally used in transport networks to accommodate relatively static traffic demand; on the other hand, traditional IP networks used to offer best-effort service, relying on hop by hop routing without QoS guarantees. The shift from traditional best-effort connectionless IP networks to multiservice MPLS-based connection-oriented networks with QoS guarantees and dynamic traffic demand imposes the challenge to solve the CBR problem with the accuracy of the offline process but the timing restriction of the online solution. CBR with more than one constraint is a well-known NP-complete problem, and therefore, the necessary computation power is unbounded. This prevents the introduction of full CBR capabilities into network devices, since they have scarce computational resources mainly devoted to packet forwarding. Moreover, different implementations of CBR algorithms lead to the impossibility of fulfilling network-wide TE objectives.

The RMA is a routing and signalling peer node in the network, but avoiding traffic forwarding. Also, it is assumed that in order to fulfil its main objective, the RMA has enough computation power to solve the CBR problem in near real time. This entity, while enabling a Control Plane-based provisioning, can be used as a complementary TE tool by management applications, using its interface towards the Management Plane.

In the tested RMA prototype, path computation can be requested using the standard RSVP-TE signalling, or by means of a request/reply protocol, in line with the aforementioned PCE WG definitions. The well-known COPS protocol [10] has been tested for this role, which additionally supports the exchange of policy information between the LSRs and the RMA. Simulation results shown in Section 4 reveal that both signalling alternatives are feasible. Further explanation of the RMA architecture can be found in [12].

2.2.4 Element Manager

This module is composed by the EMS (Element Manager System), MLSN (Multi Layer Subnetwork) and ME (Managed Element) managers, developed by the afore-mentioned BSc projects. Note that these modules are technology-independent, while the ME module implements an extension API that is used by the Device Manager (DM), which actually implements the technology-specific communication with managed elements.

The DM is an extension of the system presented in [13]. Its functionality is tailored for IP/MPLS networks. Besides the existing Cisco-specific parser, support for standard MPLS-MIB is being built in the DM. Also, the Linux nodes are being augmented with an MPLS-MIB compliant SNMP Agent using Net-SNMP extension modules [14].

3 Related Work

MPLS augments traditional best-effort IP with traffic engineering capabilities, adding functionality to network devices (i.e. constraint-based path computation), which need ever growing resources to cope with this complexity in terms of processing time. Finding optimal, or even feasible network paths that meet certain constraints is a challenging task for head-end routers. Some traffic engineering systems have been proposed to assist network devices in path computation and resource assignment in MPLS networks. Their relevant characteristics are briefly described hereafter.

RATES [15] presents a component-based, expandable management architecture, comprised of the following major components:

- Explicit Route computation.
- COPS Server for communication with head-end routers.
- Data Repository.
- Network Topology and State Discovery component, running OSPF.

Some of the ideas presented by this proposal have been considered in the design of GERMINA; in particular the usage of OSPF to gather network information is an important contribution.

Wise<TE> Traffic Engineering server for MPLS networks [16], presents a complete analysis of the caveats of IP-based traffic engineering, the limitations of CBR running on network nodes and other issues, but reaches the limited conclusion that the solution is an offline traffic engineering tool. Wise <TE> major components are the following:

- Traffic Measurement and Analysis Server, which performs collection, characterization and analysis of traffic information.
- Routing Advisor for Traffic Engineering, a planning tool with CBR capabilities.
- Resource Monitoring Server, a topology and configuration information repository
- Policy Server, which configure policies on network devices using vendor specific device agents.

This proposal is similar to the RATES system; its major contribution regarding the GERMINA architecture is, again, the dynamic gathering of network states to feed the traffic engineering database.

Both proposal are founded over a CORBA distributed environment, and present similar ideas. Based on previous work, the design of a new system is advised to be component-based, with monitoring capabilities, and with participation in the IGP process. The communication with the network devices is largely dependent on specific routers capabilities; anyhow, COPS is widely deployed and is a feasible choice, as shown in Section 4. Path computation is non-standard by nature, and each proposal implements algorithms that claim to fulfil certain performance objectives. A new proposal shall be flexible in this respect to support different capabilities and do not be restricted by computational resources.

4   Proof of Concept Results

GERMINA is a growing system composed of several disjoint modules tight together by a common CORBA framework. Functional testing of the Provisioning and Inventory Components have been conducted; results can be found in [7] and [8].

Stimulating results have been obtained by simulation of the RMA component, using ns-2 simulator [16] extensions for RSVP-TE [18] and COPS [20], with Waxman topologies generated by the BRITE tool [18]. Fig. 2. shows the summarizing results for the establishment of LSPs using the Management Plane, the Control Plane and the hybrid (i.e. the RMA) approaches.

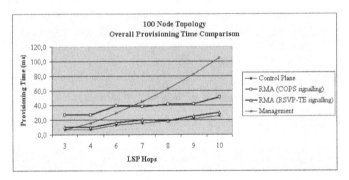

Fig. 2. LSP Setup Time as a function of LSP hops

Provisioning time for the Management Plane and the Control Plane approaches impose upper and lower bounds respectively. The RMA using RSVP-TE signalling shows a performance very close to the Control Plane lower bound, while the RMA with COPS signalling timing shows a slight degradation due to the complexities added by the session establishment between the ingress LSRs and the COPS server (i.e. the RMA). Note that in this case, bidirectional LSP setup is achieved, so there is a trade-off between a slower response time and an enhanced functionality.

Field evaluation of the GERMINA management system is being undertaken in a trial network. The core is composed by standard commercial MPLS routers, while the aggregation and access is comprised of Linux-based routers, as shown in Fig. 3. .

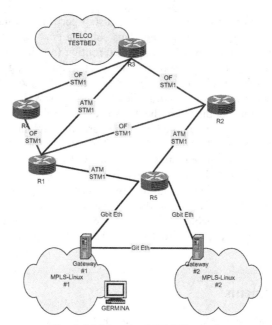

Fig. 3. Multiservice MPLS Network

The MPLS network runs the OSPF-TE routing protocol [21] and uses RSVP-TE signalling for the establishment of Traffic Engineered LSPs (TE-LSPs). The GERMINA management system is located in one of the access "clouds" and gathers network information using a Management Virtual Private Network built over the MPLS infrastructure. A set of tools for traffic generation and monitoring are used to emulate the users' behaviour, modelled using real network traces gathered from strategic nodes of the public IP network. Gathering and processing network traces for user characterization is one of the interesting outcomes of the project, motivating the usage of data warehousing and statistical techniques to cope with the enormous amount of data needed to be handled.

As seen on the picture, the infrastructure comprises point to point STM-1 optical links, ATM switched STM-1 and Gigabit Ethernet links. The core is composed of Cisco 7206 VXR routers and the Gateways are Intel servers equipped with PRO/1000 MF Dual Port adapters, running MPLS-Linux [22] and Quagga routing software [23]. As mentioned above, Net-SNMP tools are installed in the Linux nodes, which implement a subset of the MPLS-MIBs.

5 Conclusions and Future Work

This paper presents an ongoing project entailed to provide a solution for the O&M challenge imposed by new technologies and services being deployed in Service Providers' networks. The GERMINA management system provides alternatives for service provisioning using a standard distributed, component-based framework augmented with a hybrid concept, the RMA, which takes advantage of the MPLS Control Plane.

Simulation results for the RMA show promising performance in comparison with "pure" Control Plane LSP provisioning. This solution enables a global optimisation of network resources by means of an improved routing function outside LSRs, which also offload router processing, boosting the packet forwarding functionality. The RMA architecture is entirely based on standard protocols, and enables network administrators to control the routing strategies (i.e. network-wide TE objectives). Further validation is being conducted in the described trial testbed.

Future work involves the completion of the partially implement modules, conception of a mechanism to define network-wide TE policies for the CBR module, and trial of exact algorithms and heuristics to solve the CBR problem near real time.

An evolution towards a Service Oriented Architecture (SOA) is being evaluated as a foundation for the management building blocks, while aggregation and customer-specific issues shall be considered in the trial network. For example the usage of IP DSLAMs and multicast-enabled Ethernet aggregation over MPLS Virtual Private LAN Services (VPLS) and/or Virtual Private LAN (VLAN) are topics that will be evaluated.

Acknowledgements

The authors would like to express their gratitude to their co-workers in the Metropolitan Multiservice Network with MPLS over optical transport project at UdelaR, Uruguay, for their support and valuable comments.

References

1. MINA Academic Group: http://www.fing.edu.uy/inco/grupos/mina. Last visited: August 2005.
2. L. Raptis, G. Hatzilias, F. Karayannis, K. Vaxevanakis, E. Grampín, "An integrated network management approach for managing hybrid IP and WDM networks". *IEEE Network*, Volume: 17, Issue: 3, May-June 2003 Pages:37 – 43.
3. D. Allan, T.D. Nadeau, "A Framework for MPLS Operations and Management OAM", Internet Draft <draft-ietf-mpls-oam-frmwk-03.txt>. Work in Progress. Expiration Date: August 2005.
4. JacORB: http://www.jacorb.org. Last visited: August 2005.
5. TMF 608, "Multi-Technology Network Management Information Agreement NML-EML Interface." Version 2.0. TM Forum Approved, October 2001.
6. "Connection and Service Management Information Model (CaSMIM)" TMF 508,605 and 807 Public Evaluation, June 2001, Version 1.5
7. M. Giachino, C. González, J. Visca. "Component-based Network Management". BSc Graduation Project, April 2004. Universidad de la República, Uruguay.
8. M. Bartesaghi, B. Fagalde, F. Zubía. "Inventory Management". BSc Graduation Project, March 2004. Universidad de la República, Uruguay.
9. D. Awduche et al, "RSVP-TE: Extensions to RSVP for LSP Tunnels". RFC 3209, December 2001.
10. IETF Path Computation Element (pce) Working Group. http://www.ietf.org/html.charters/pce-charter.html
11. D. Durham, J. Boyle, R. Cohen, S. Herzog, R. Rajan, A. Sastry, "The COPS (Common Open Policy Service) Protocol". RFC 2748, January 2000.

12. E. Grampín, J. Serrat. "Cooperation of Control and Management Plane for Provisioning in MPLS Networks". IM 2005 Conference, Nice, France. May 15-19, 2005.
13. E. Grampín, J. Baliosian, J. Serrat. "Extensible, Transactional Architecture for IP Connectivity Management". 3rd IEEE Latin American Network Operations and Management Symposium (LANOMS'2003). Iguaçu Falls, Brazil - Setiembre 4 - 6, 2003.
14. Net-SNMP: http://www.net-snmp.org/. Last visited: August 2005.
15. P. Aukia, M. Kodialam, P.V. Koppol, T.V. Lakshman, H. Sarin, B. Suter, "RATES: a server for MPLS traffic engineering". *IEEE Network*, Volume: 14, Issue: 2, March-April 2000 Pages:34 - 41.
16. T.S. Choi, S.H. Yoon, H.S. Chung, C.H. Kim, J.S. Park, B.J. Lee, T.S. Jeong, "Wise<TE>: traffic engineering server for a large-scale MPLS-based IP network". Network Operations and Management Symposium, IEEE/IFIP NOMS 2002, 15-19 April 2002.
17. The Network Simulator - ns-2: http://www.isi.edu/nsnam/ns/. Last visited: August 2005.
18. A. Medina, A. Lakhina, I. Matta, J. Byers, "BRITE: Boston university Representative Internet Topology gEnerator", http://www.cs.bu.edu/brite/ Last visited: August 2005.
19. C. Callegari, F. Vitucci, "RSVP-TE/ns network simulator", http://netgroup-serv.iet.unipi.it/rsvp-te_ns/. Last visited: July 2005.
20. Sayenko, T. Lahnalampi, "COPS for the NS-2 simulator", http://www.cc.jyu.fi/~sayenko/pages/en/projects.htm. Last visited: July 2005.
21. D. Katz, K. Kompella, D. Yeung, "Traffic Engineering Extensions to OSPF Version 2", RFC 3630, September 2003.
22. MPLS for Linux project: http://sourceforge.net/projects/mpls-linux/. Last visited: July 2005.
23. Quagga Routing Software Suite: http://www.quagga.net/. Last visited: July 2005.

Performance of Traffic Engineering in Operational IP Networks – An Experimental Study*

Anders Gunnar, Henrik Abrahamsson, and Mattias Söderqvist

SICS – Swedish Institute of Computer Science
{aeg,henrik,mso}@sics.se

Abstract. Today, the main alternative for intra-domain traffic engineering in IP networks is to use different methods for setting the weights (and so decide upon the shortest-paths) in the routing protocols OSPF and IS-IS. In this paper we study how traffic engineering perform in real networks. We analyse different weight-setting methods and compare performance with the optimal solution given by a multi-commodity flow optimization problem. Further, we investigate their robustness in terms of how well they manage to cope with estimated traffic matrix data. For the evaluation we have access to network topology and traffic data from an operational IP network.

1 Introduction

For a network operator it is important to tune the network in order to accommodate more traffic and meet service level agreements (SLAs) made with their customers. In addition, as new bandwidth demanding and also delay and loss sensitive services are introduced it will be even more important for the operator to manage the traffic situation in the network. This process of managing the traffic is often referred to as *traffic engineering*. The aim is to use the network resources as efficiently as possible and to avoid congestion; *i.e.* deviate traffic from highly utilized links to less utilized links.

In this paper we investigate performance of traffic engineering in operational networks. To make the investigation more balanced we use two traffic engineering methods. The results are compared to the optimal routing obtained from multi commodity flow optimization and the inverse capacity weight setting recommended by Cisco. In order to optimize the routing an estimate of the traffic situation in the network is needed. The traffic situation can be captured in a traffic matrix. The entries in the traffic matrix represent the amount of traffic sent between each source destination pair in the network. However, since routers often lack functionality to measure the traffic matrix directly operators

* This work was supported by Vinnova via the EvaluNet project and by the Winternet program which is funded by the Swedish Foundation for Strategic Research. The work was also supported by the European commission via the Ambient Networks project

T. Magedanz, E.R.M. Madeira, and P. Dini (Eds.): IPOM 2005, LNCS 3751, pp. 202–211, 2005.

are forced to estimate it from other available data. We use two well known traffic matrix estimation methods to investigate how traffic engineering perform when subjected to estimated traffic matrices.

For the evaluation we have access to a full traffic matrix as well as network topology obtained from direct measurements in a commercial IP network. Previous work have shown that traffic engineering enables the network operator to accommodate substantially more traffic in the network. However, the evaluations have been performed on synthetic data or only partial traffic matrices obtained from an operational IP network.

Our focus is not on the actual methods we use in the study but in how they perform in a real network with real traffic demands. In addition, we study the interplay between traffic estimation and an application of the estimate, *i.e.* traffic engineering. Hence, we only give a brief description of the methods we use and the interested reader should consult the references for further details.

The rest of the paper is organized as follows. Section 2 give a short description of traffic engineering in IP networks. We also discuss related work on the subject. The experiments together with a short description of traffic matrix estimation is given in Section 3. The evaluation is described in Section 4. Finally we make some concluding remarks about our findings and discuss future work.

2   Traffic Engineering in IP Networks

Traffic engineering encompasses performance evaluation and performance optimization of operational networks. An important goal is to avoid congestion in the network and to make better use of available network resources by adapting the routing to the current traffic situation.

The two most common intra-domain routing protocols today are OSPF (Open Shortest Path First) and IS-IS (Intermediate System to Intermediate System). They are both link-state protocols and the routing decisions are based on link costs and a shortest (least-cost) path calculation. With the equal-cost multi-path (ECMP) extension to the routing protocols the traffic can also be distributed over several paths that have the same cost.

These routing protocols are designed to be simple and robust rather than to optimize the resource usage. They do not by themselves consider network utilization and do not always make good use of network resources. The traffic is routed on the shortest path through the network even if the shortest path is overloaded and there exist alternative paths. It is up to the operator to find a set of link costs (weights) that is best suited for the current traffic situation and avoids congestion in the network.

The traffic engineering process is illustrated in Figure 1. The first step is to collect the necessary information about network topology and the current traffic situation. Most traffic engineering methods need as input a traffic matrix describing the demand between each pair of nodes in the network. But today the support in routers for measuring the traffic matrix is limited. Instead, an often suggested approach is to estimate the traffic matrix from link loads and routing

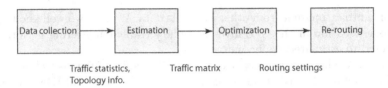

Fig. 1. The traffic engineering process

information [5, 6, 13]. Link loads are readily obtained using the Simple Network Management Protocol (SNMP) and routing information is available from OSPF or IS-IS link-state updates.

The traffic matrix is then used as input to the routing optimization step, and the optimized parameters are finally used to update the current routing. In this study this means that the traffic matrix is used together with heuristic search methods to find the best set of links weights.

2.1   Optimal Routing

The general problem of finding the best way to route traffic through a network can be mathematically formulated as a multi-commodity flow (MCF) optimization problem (see, e.g., [1, 3, 7]). The network is then modeled as a graph. The problem consists of routing the traffic, given by a demand matrix, in the graph with given link capacities while minimizing a cost function. This can be formulated and solved as a linear program.

How the traffic is distributed in the network very much depends on the objectives expressed in the cost function. Since one of the main purposes with traffic engineering is to avoid congestion a reasonable objective would be to minimize the maximum link utilization in the network. Another often proposed objective function is described by Fortz and Thorup [3]. Here the sum of the cost over all links is considered and a piece-wise linear increasing cost function is applied to the flow on each link. The basic idea is that the cost should depend on the utilization of a link and that it should be cheap to use a link with small utilization while using a link that approaches 100% utilization should be heavily penalized. The characteristics of the cost function is shown in Figure 2.

Though the solution given by the linear program is the optimal routing the method is in general not used directly for routing in operational IP networks. First, the method is inherently centralized. And also, since the solution requires flows to be arbitrary split among several paths towards the destination it would require modifications to the forwarding mechanisms that is used today [1]. In this paper we focus on the legacy routing protocols OSPF and IS-IS. The optimal routing will be used for comparison only since it constitutes a lower bound for the performance of legacy routing mechanisms.

Unfortunately, when taking the restrictions of shortest-paths or equal-cost multipaths in the OSPF and IS-IS protocols into consideration, the problem of finding the optimal routing becomes much harder. The problem of finding

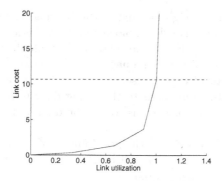

Fig. 2. Cost function for load on a link

weights that optimizes the routing is NP-hard [3, 7]. This means that one usually has to rely on heuristic methods to find the set of weights.

2.2   Heuristic Search Methods

An often proposed method to determine the best set of link weights is to use local search heuristics [3, 4, 8]. Given network topology, link capacities and the demand matrix the heuristics evaluate points in a search space, where a point is represented by a set of weights. A neighbor to a point is another set of weights produced by changing the value of one or more weights from the first point. In the heuristics, different neighbors are produced and the cost of each one is calculated using a cost function. From each heuristic the neighbor with the best cost is the one that will be the output. In this study we have selected two heuristics:

- Local search (Fortz and Thorup [3])
- Strictly descending search (Ramakrishnan and Rodrigues [8])

Both heuristics have been studied for random topologies and synthetic traffic demands by Söderqvist [10]. As objective function we use the cost function by Fortz and Thorup [3] mentioned in section 2.1.

2.3   Related Work

With the prospect of better utilizing available network resources and optimizing traffic performance, a lot of research has been done in the area of traffic engineering. The general principles and requirements for traffic engineering are described in RFC 3272 [2] produced by the IETF Internet Traffic Engineering working group.

Many researchers use multi-commodity flow models for traffic engineering. The book by Pióro and Medhi [7] gives a comprehensive description of design models and optimizations methods for communication networks, including networks with shortest-path routing.

The performance of weight-setting methods using search heuristics has been investigated with real network topologies and synthetic data or partial traffic matrices in [3, 4, 9, 11]. Fortz and Thorup [3] evaluate their search heuristic using a proposed AT&T backbone network and demands projected from measurements. Sridharan *et al.* [11] use a heuristic to allocate routing prefixes to equal-cost multi-paths and evaluate this using data from the Sprint backbone network. An alternative approach is to use the dual of a linear program to find a weight setting [12].

Roughan *et al.* [9] investigate the performance of traffic engineering methods with estimated traffic matrices. However, the authors use partial traffic matrices and one weight setting method only.

3 Methodology

This section describes the methodology in this study. First we describe the performance metrics for the experiments followed by a discussion on how the experiments are conducted. Finally we give a short introduction to the traffic matrix estimation methods used in this paper.

3.1 Evaluation Metrics

Since the main objective of traffic engineering is to avoid congestion one natural metric of performance is maximum link utilization in the network. The utilization u_a of link a is defined as:

$$u_a = \frac{l_a}{c_a} \qquad (1)$$

where the load on link a is denoted l_a and c_a is the capacity of the link. However, maximum link utilization only reflect one link in the network. To quantify performance for the routing where the whole network is taken into account we define the normalized cost:

$$\Phi^* = \frac{\Phi}{\Phi_{norm}}. \qquad (2)$$

Here Φ is the cost function from Section 2.1 and Φ_{norm} is a normalization factor such that the normalized cost is comparable between different network topologies. Further details about the normalized cost can be found in [10].

3.2 Experimental Setup

In this paper we use an experimental approach to address the problem. We use a unique data set of complete traffic matrices and topology from a commercial IP network operator to simulate the effects of different weight settings for OSPF/IS-IS routing.

A measured traffic matrix, *i.e.* a traffic matrix without errors, together with the network topology is provided as input to the weight optimization. The output from the optimization is a new set of weights which we use to calculate the new routing. Finally, the measured traffic matrix is applied to the new routing in order to determine link utilization and calculate the normalized cost.

To obtain an estimated traffic matrix we simulate the routing with inverse capacity routing. The link loads obtained by applying the measured traffic matrix is then used to find an estimate of the traffic matrix. The estimated traffic matrix is used as input to the weight optimization algorithm. Finally, the optimized links weights are used to calculate the links loads by applying the original measured traffic matrix.

The optimal solution to the routing problem discussed in section 2.1 will serve as a benchmark for our experiments with the search heuristics. In addition, the routing from the inverse capacity weight setting is also included for comparison as it is often used by network operators and is the recommended weight setting by Cisco [3].

3.3   Traffic Matrix Estimation

The traffic matrix estimation problem has been addressed by many researchers before (e.g. [5, 6, 13]). In this study we focus on two estimation methods.

- Simple Gravity method
- Entropy method

The simple gravity method is based on the assumption that traffic between source node s and destination node d is proportional to the total amount of traffic sent by s and total amount of traffic destined to d. The strength of this method lies in its simplicity. However, the method is also known to be unaccurate in some situations [5].

A different approach to obtain the traffic matrix is to estimate it from link loads and routing information [6, 13]. Link loads are readily obtained using SNMP and routing is available from OSPF or IS-IS link-state updates. This approach often leads to an ill-posed estimation problem since operational IP networks typically have many more node pairs (entries in the traffic matrix) than links. In order to add more constraints to the problem additional information must be added. This information is usually in the form of some assumption made about the traffic matrix. The entropy method [13] minimizes the Kullback-Leibler distance between the estimate and a prior guess of the traffic demands. With the entropy method it is possible to obtain an accurate estimate of the traffic matrix (cf. [5, 13]). In our experiments we use the gravity method to produce a prior.

By choosing one accurate and one less accurate method we are able to make a more balanced evaluation of how estimation errors influence the performance of traffic engineering subjected to estimated traffic demands.

4 Results

In this section we present the results obtained from our experiments. For the evaluation we used network topologies and traffic matrices obtained from a global MPLS-enabled IP network. From the data we isolated the European and the American subnetworks in order to obtain networks of manageable size but still carry large traffic demands. In addition, we obtain two networks with slightly different characteristics. More details about the networks and traffic demands can be found in Gunnar *et al.* [5]. However, it might be interesting to mention that the European network has 12 nodes and 40 links and the American network has 25 nodes and 112 links.

As previously mentioned we use two measures of performance, the normalized cost function introduced by Fortz and Thorup [3] and maximum link utilization in the network. The results are plotted for the following methods:

- **Opt**, the optimal solution to the general routing problem. Included for comparison since it is a lower bound for the other methods.
- **InvCap**, sets the weight inversely proportional to the capacity of the link. Like Opt this method is included as a benchmark as it is the default setting recommended by Cisco.
- **FT**, the search heuristic proposed by Fortz and Thorup [3] starting from a random weight setting.
- **RR**, the search heuristic proposed by Ramakrishnan and Rodrigues [8].

For each topology the algorithms were run with different scaling on the traffic demands. The scalings were obtained by multiplying the traffic demand matrix with a scalar. All algorithms except InvCap use different weight settings for different scalings. In the results both cost and max utilization are presented for each topology and for each scaling. For all algorithms except OPT the cost and the max utilization is computed using the same weight setting. But for OPT the cost and the max utilization are computed independently, using different objective functions.

4.1 Experiments with Measured Traffic Matrices

Figure 3 shows the normalized cost and maximum link utilization for the American network and for both search heuristics. The plot shows that both search heuristics are close to the optimal routing given by the linear programming model. However, in the European network the results are somewhat different as Figure 4 reveals. Both heuristics improve performance compared to inverse capacity weight setting but neither of them are close to the optimal routing.

In comparison with previous studies we see that our findings confirm the results of Fortz and Thorup [3] who use a real network topology and a partial traffic matrix derived from Netflow measurements as well as synthetic data. Söderqvist [10] use synthetic topologies and traffic demands with power-law properties to show that optimizing weights improve network performance considerably compared to inverse capacity weight setting.

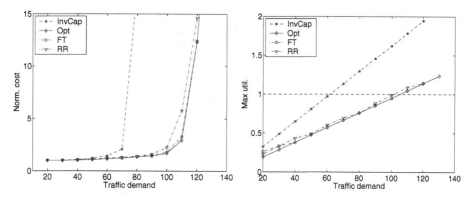

Fig. 3. Normalized cost (left) and maximum link utilization (right) for the American network

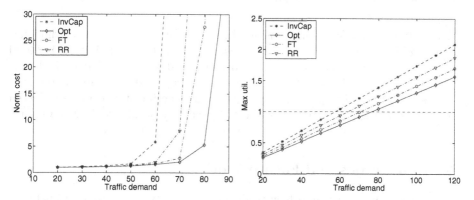

Fig. 4. Normalized cost (left) and maximum link utilization (right) for the European network

4.2   Optimizing Weights Using Estimated Traffic Demands

A somewhat controversial assumption made in the previous section is that an exact measure of the traffic matrix is available. In this section we investigate how the search heuristics perform when they are subjected to estimated traffic demands.

We focus on two well known estimation methods. The gravity method and the entropy method. Both methods have been evaluated on the data set we use in this study [5]. The simple gravity methods was shown to give a surprisingly accurate estimate in the European network despite its simplicity. In the American network, on the other hand, the gravity methods failed to give an accurate estimate of the traffic demands due to violation of the gravity assumption. The more sophisticated entropy method produced an accurate estimate for both the European and the American networks.

In the plot to the left in Figure 5 we have plotted normalized cost as a function of traffic demands for the local search heuristic in the American network. The

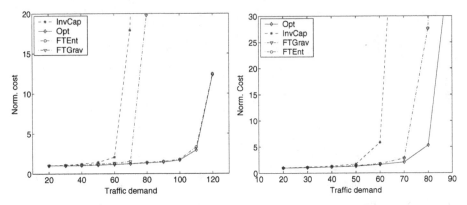

Fig. 5. Normalized cost for the American network (left) and the European network (right) using estimated traffic matrices

plot indicates that estimation using the more advanced entropy method has a negligible effect on performance. However, when the optimization is based on the less accurate gravity model performance is degraded considerably. In the European network (Figure 5 right), where the gravity model is more accurate, the heuristics have similar performance. The same experiment has been conducted using descending search producing similar results as local search. But we have omitted the plots for descending search due to space limitations.

5   Conclusions and Future Work

This paper has investigated how traffic engineering performs in a real network with real traffic demands from a commercial IP network operator. The traffic engineering methods are based on search heuristics for weight settings in link state routing. From the study we concluded that both search heuristics are able to find weight settings which are able to accommodate substantially more traffic in the network than the default inverse capacity weight setting and come close in performance to the optimal solution of the routing problem. In addition, we study the performance of the traffic engineering using estimated traffic demands. We investigate two traffic matrix estimation methods. One simple and one which is more sophisticated and accurate. Our observations indicate that when the optimized weight setting using the estimated traffic matrix from the accurate entropy method is applied to the real traffic demands performance is only degraded marginally. But for the less accurate gravity model performance was degraded significantly in some cases. However, still an improvement compared to the inverse capacity weight setting recommended by Cisco.

Our findings confirm the results form previous studies using partial traffic demands derived from flow measurements or synthetic data [3, 9, 10].

This study has focused on a static traffic matrix. In the future we intend to investigate how the weight setting can be designed to be robust in order to cope with a changing traffic situation in the network.

Acknowledgments

This paper describes work undertaken in the context of the Ambient Networks - Information Society Technologies project, which is partially funded by the Commission of the European Union. The views and conclusions contained herein are those of the authors and should not be interpreted as necessarily representing the Ambient Networks Project.

References

1. H. Abrahamsson, J. Alonso, B. Ahlgren, A. Andersson, and P. Kreuger. A Multi Path Routing Algorithm for IP Networks Based on Flow Optimisation. In *Proceedings of QofIS 2002*, pages 135–144, Zürich, Switzerland, Oct 2002.
2. D. Awduche, A. Chiu, A. Elwalid, I. Widjaja, and X. Xiao. Overview and principles of Internet Traffic Engineering. Internet RFC 3272, May 2002.
3. B. Fortz and M. Thorup. Internet Traffic Engineering by Optimizing OSPF Weights. In *Proceedings IEEE INFOCOM 2000*, pages 519–528, Israel, March 2000.
4. B. Fortz and M. Thorup. Optimizing OSPF/IS-IS weights in a changing world. *IEEE Journal on Selected Areas in Communications*, 20(4):756–767, May 2002.
5. A. Gunnar, M. Johansson, and T. Telkamp. Traffic Matrix Estimation on a Large IP Backbone - a Comparison on Real Data. In *Proceedings of ACM Internet Measurement Conference*, Taormina, Sicily, Italy, October 2004.
6. A. Medina, N. Taft, K. Salamatian, S. Bhattacharyya, and C. Diot. Traffic matrix estimation: Existing techniques and new directions. In *Proc. ACM SIGCOMM*, Pittsburg, USA, August 2002.
7. M. Pióro and D. Medhi. *Routing, Flow, and Capacity Design in Commmunication and Computer Networks*. Morgan Kaufmann, 2004.
8. K.G. Ramakrishnan and M.A. Rodrigues. Optimal routing in shortest path data networks. *Lucent Bell Labs Technical Journal*, 6(1), 2001.
9. M. Roughan, M. Thorup, and Y. Zhang. Traffic Engineering with Estimated Traffic Matrices. In *Proceedings ACM Internet Measurement Conference*, Miami Beach, Florida, USA, October 2003.
10. M. Söderqvist. Search Heuristics for Load Balancing in IP-networks. Technical Report T2005:04, SICS – Swedish Institute of Computer Science, March 2005.
11. A. Sridharan, R. Guerin, and C. Diot. Achieving Near-Optimal Traffic Engineering Solutions for Current OSPF/IS-IS Networks. In *Proceedings of IEEE INFOCOM 2003*, San Francisco, USA, mars 2003.
12. Y. Wang, Z. Wang, and L. Zhang. Internet traffic engineering without full mesh overlaying. In *Proceedings of IEEE INFOCOM 2001*, Anchorage, Alaska, 2001.
13. Y. Zhang, M. Roughan, C. Lund, and D. Donoho. An information-theoretic approach to traffic matrix estimation. In *Proc. ACM SIGCOMM*, Karlsruhe, Germany, August 2003.

Author Index